Advances in Soil Science

Advances in Soil Science

B.A. Stewart, Editor

Advances in Soil Science

Volume 14

Edited by B.A. Stewart

With Contributions by
J.B. Harsh, J.M. McCray, L.K. Porter, J.S. Schepers,
N. Senesi, S.J. Smith, M.E. Sumner, S.J. Traina, and Shihe Xu

With 74 Illustrations

Springer-Verlag
New York Berlin Heidelberg
London Paris Tokyo Hong Kong

B.A. Stewart
USDA Conservation & Production Research Laboratory
Bushland, Texas 79012, U.S.A.

ISSN: 0176-9340

Printed on acid-free paper.

Typeset by Publishers Service, Bozeman, Montana.

9 8 7 6 5 4 3 2 1

ISBN-13:978-1-4612-7978-5 e-ISBN-13:978-1-4612-3356-5
DOI: 10.1007/978-1-4612-3356-5

Preface

The study of soils today has taken on increased importance because a rapidly expanding population is placing demands on the soil never before experienced. This has led to an increase in land degradation. Land degradation is one of the most severe problems facing mankind. Volume 11 of *Advances in Soil Science* was devoted entirely to this critical area of soil science.

From the beginning of agriculture until about 1950, increased food production came almost entirely from expanding the cropland base. Since 1950, however, the yield per unit of land area for major crops has increased dramatically. Much of the increase in yields was because of increased inputs of energy. Between 1950 and 1985, the farm tractor fleet quadrupled, world irrigated area tripled, and use of fertilizer increased ninefold. Between 1950 and 1985, the total energy used in world agriculture increased 6.9 times.

Until recently, sustainability was seldom, if ever, mentioned in agricultural literature. Now, it is one of the most widely used terms. The high costs of irrigation development, escalating energy costs during the 1970s, public concern over potential negative impacts of fertilizer and pesticides on water supplies, soil erosion, soil compaction and salinity problems, and other concerns have caused many people to question whether many of the present agriculture systems can be sustained. As a result, soil science is beginning to focus more on sustaining the resource base. The productivity level of an agricultural soil at any point in time is the result of the interaction of degradative processes and conservation/reclamation practices that occur simultaneously. A sustainable system is any system in which the benefits from the soil conservation practices are equal to or greater than the negative effects of the soil degradative processes.

This series, *Advances in Soil Science*, was established to provide a forum for leading scientists to analyze and summarize the available scientific information on a subject, assessing its importance and identifying additional research needs. This goal seems even more appropriate today than in 1982 when the idea of the series was formulated. Much has been learned about our soil resources. The principles learned and the technology developed need to be used to increase food production and sustain the productivity of the resource base. *Advances in Soil*

Science fills a gap between the scientific journal and the comprehensive reference books. Scientists can delve in depth into a particular subject relating to soil science. Contributors are asked to develop and identify principles that have practical applications to both developing and developed agricultures.

Advances in Soil Science is international in scope and covers all subjects relating to soil science. This volume continues that format. Although we consider our audience to be primarily scientists and students of soil science, the series provides technical information to anyone interested in our natural resources and man's influence on these resources. Research in the future will focus on systems that are resource efficient and environmentally sound. The need to optimize crop production while conserving the resource base has never been greater.

The quick acceptance of *Advances in Soil Science* by both authors and readers has been very gratifying and confirms our perception that a need did exist for a medium to publish soil science reviews. I want to thank the authors for their excellent contributions and cooperation. I also want to thank members of the Editorial Board for their help in selecting such competent authors and the Springer-Verlag staff for their kind assistance and counsel. Last, and most important, I want to thank the readers for their acceptance and use of *Advances in Soil Science*.

B.A. Stewart

Contents

Contributors

J.B. Harsh, Department of Agronomy and Soils, Washington State University, Pullman, Washington 99164, U.S.A.

J.M. McCray, Research Department, U.S. Sugar Corporation, Clewiston, Florida 33440, U.S.A.

L.K. Porter, U.S. Department of Agriculture, Agricultural Research Service, Ft. Collins, Colorado 80522, U.S.A.

J.S. Schepers, U.S. Department of Agriculture, Agricultural Research Service, Lincoln, Nebraska 68583, U.S.A.

N. Senesi, Istituto di Chimica Agraria, Faculty of Agriculture, University of Bari, 70126 Bari, Italy.

S.J. Smith, U.S. Department of Agriculture, Agricultural Research Service, Durant, Oklahoma 74702, U.S.A.

M.E. Sumner, Department of Agronomy, The University of Georgia College of Agriculture, Athens, Georgia 30602, U.S.A.

S.J. Traina, Department of Agronomy, Ohio State University, Columbus, Ohio 43210, U.S.A.

Shihe Xu, Department of Agronomy and Soils, Washington State University, Pullman, Washington 99164, U.S.A.

Assessing and Managing Agricultural Nitrogen Losses to the Environment[1]

S.J. Smith, J.S. Schepers, and L.K. Porter

[1]Contribution from the U.S. Department of Agriculture, Agricultural Research Service, Southern Plains and Northern Plains Areas.

1990 by Springer-Verlag New York Inc.
Advances in Soil Science, Volume 14

I. Introduction

Major concern continues over the potentially harmful environmental impacts associated with excessive agricultural nitrogen (N) losses. The concern is about losses in both water and atmospheric quality (Gilliam et al., 1985; Keeney, 1982a; Nielsen and MacDonald, 1978). Water quality concern has focused mainly on groundwater nitrate levels considered health-threatening and accelerated eutrophication of surface waters, whereas atmospheric concern focuses on increased emission of nitrous oxide associated with N fertilizer application. Concern about increased nitrous oxide emission stems from its implication in stratospheric reactions which contribute to depletion of the ozone that protects the earth from ultraviolet radiation.

Nitrogen, an essential major plant nutrient, represents the mineral fertilizer most applied to agricultural lands. This is because available soil N supplies are generally inadequate for optimum crop production. In the ammonium form, N is fairly immobile in soil. Under most conditions, however, ammonium is converted biologically to nitrate, which readily moves with the soil water. That nitrate not used by the crop has the potential to move from the soil to groundwater, streams, and impoundments. In addition, some nitrate denitrifies and passes to the atmosphere in a gaseous form. Dissipation to the atmosphere is also possible with ammonium, especially under alkaline soil conditions. Moreover, considerable particulate-soil organic nitrogen is moved through erosional processes. Overall, then, agricultural nitrogen can be lost to the environment in liquid, gaseous, or solid form.

This chapter considers pertinent factors and approaches that are useful in assessing and managing agricultural N losses to the environment. The treatment here covers work done primarily over the past 10 to 15 years. For earlier information, readers are referred to publications by Stanford et al. (1970), Viets and Hageman (1971), and Wadleigh (1968). Basically, however, the solution to minimizing agricultural N losses to the environment remains the same—that is, increasing fertilizer N efficiency. Not to be overlooked are the environmental benefits N fertilizers provide. By increasing crop cover, they can reduce runoff and soil erosion. In addition, N fertilizer application to the more productive lands allows less suitable, fragile lands to be removed from cultivation.

II. Contributions to N Input

To place the topic in proper perspective, it is important to recognize that N from natural sources has always existed in soils and waters. This N has originated from geological, biological, and atmospheric sources. In fact, some soil extracts in California have been reported to contain as much as 2000 mg nitrate-N L^{-1}, primarily from geological sources (Strathouse et al., 1980). Moreover, it is proper to note that soils in their virgin states contained considerably higher N contents than they do today, under cropped, fertilized conditions. For example, in a study involving eight major U.S. agricultural soils (Smith and Young, 1975), total N contents of the virgin, level, surface horizons contained 760 to 4110 mg Total Kjeldahl Nitrogen kg^{-1}, on the average, one-third more than their N fertilized, cultivated counterparts that had been farmed 15 to 70+ years. Consequently, from an environmental standpoint, the concern is not that there is too much soil N, but that too much N is present in the wrong form (e.g., nitrate) and place (i.e., water supplies).

Annually, only about 3% or less of the total soil N is converted to a soluble, plant-available form (Bremner, 1965). For optimum crop production, this means supplemental N sources are often required. Such N sources include fertilizer, feedlot waste and manure, biological fixation, irrigation water, and wet–dry atmospheric deposition. The goal is to see that the crop uses as much available N as is prudently possible from the soil and supplemental sources, thereby minimizing N losses to runoff, leaching, denitrification, volatilization, deposition, and so forth. Considering that the current efficiency of the supplemental N sources is often <50%, appreciable room remains for improvement. A prime mission of agricultural N researchers is to develop workable technology that will enhance this efficiency.

III. N Loss Processes

A. Wind Erosion

Legg and Meisinger (1982) estimated that 4.5 million metric tons of U.S. cropland N are lost annually to wind and water movement. Twenty percent, or 0.9 million tons, was attributed to wind. As with water, erosion by wind is a selective process, removing finer particles containing a disproportionately greater amount of N. The degree of N enrichment in the windborne soil material relative to that of the contributing soil material can be expressed as an enrichment ratio. Sparse data exist on N enrichment ratios for wind erosion, but would appear to be about 2 and 3 for organic and inorganic N, respectively (Hagen and Lyles, 1985; Zobeck and Fryrear, 1986). More exacting prediction techniques are under development (J.R. Williams, 1989, personal communication) and, until this is accomplished, these values can be incorporated with the amount of wind erosion to approximate N loss. The amount of wind erosion can be calculated by the equation (Woodruff and Siddoway, 1965)

$$WE = f(I, WC, WK, WL, VE)$$

where WE is the wind erosion amount, I is the soil erodibility index, WC is the climatic factor, WK is the soil-ridge roughness factor, WL is the field length along the prevailing wind direction, and VE is the quantity of small-grain equivalent vegetative cover. This equation has been modified by Cole et al. (1982) to calculate wind erosion on a daily basis.

B. Runoff

Nitrogen moves in surface runoff in both soluble and particulate forms (Sharpley et al., 1987). The dominant soluble forms are nitrate, which tends to move down the soil profile with the initial infiltrating water of a storm event, and ammonium, which tends to attach to the soil's cation exchange complex. For many agricultural watersheds, then, soluble N concentrations in runoff are fairly low and well within potable limits (i.e., 10 and 0.5 mg L^{-1}, respectively, for nitrate-N and ammonium-N). In fact, the soluble N concentrations in runoff are often less than corresponding values in rainfall. This means watersheds may act as filters in removing soluble rainfall N from runoff and suggests that there is little point in trying to predict soluble N runoff losses on the basis of soil N characteristics. Nevertheless, there are cases when high soluble N concentrations may be observed in runoff. Instances include when a soil horizon barrier (e.g., fragipan) exists in the profile, resulting in interflow (Lehman and Ahuja, 1985; Kissel et al., 1976) that reappears as surface runoff, and when a major runoff event occurs shortly after surface application of N fertilizer (Smith et al., 1988). Then, even high ammonium-N fertilizer concentrations may be observed (i.e., 10^{+} mg NL^{-1}). The situation is not long lasting, though, and often by the next runoff event soluble N runoff concentrations are close to background. In the Southern Plains, less than 5% of the applied N fertilizer tends to be lost as agricultural runoff (Kissel et al., 1976; Smith et al., 1983b). Similar observations have been made for other areas, including Iowa (Alberts et al., 1978) and Louisiana (Dunigan et al., 1976).

Particulate N (PN) losses in runoff occur primarily in the form of soil organic matter components associated with the suspended sediment. Such losses may be calculated using an N enrichment ratio (NER) approach, as described by Sharpley et al. (1985), where

$$PN = \text{Soil PN} \cdot \text{sediment concentration} \cdot \text{NER}$$

with soil units as mg kg^{-1} and sediment units as g L^{-1}. The NER is predicted by the equation (Menzel, 1980)

$$\ln(\text{NER}) = 2.00 - 0.20 \ln \text{soil loss}$$

where the units of soil loss for each runoff event are kg ha^{-1}. This approach has been shown to provide realistic particulate N loss estimates for a wide range of agricultural soils on both an event and an annual basis (Smith et al., 1986). The annual particulate N losses ranged from almost nil to 7 kg ha^{-1}.

C. Leaching

Nitrogen leaching in soils occurs primarily in the nitrate form. While some leaching of ammonium may occur in sandy soils, its leaching susceptibility is reduced by adsorption to the soil's cation exchange complex. Neither does leaching of soluble soil organic N components appear to be a major problem (Smith, 1987).

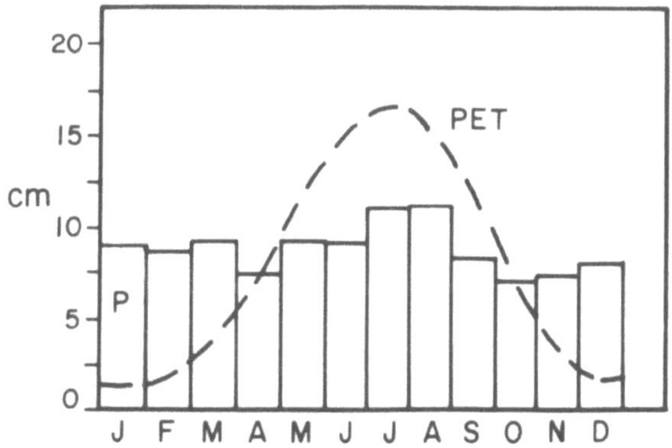

Figure 1. Precipitation and Potential EvapoTranspiration (P,PET) for Raleigh, North Carolina (Smith and Cassel, 1990).

Various approaches exist for estimating nitrate leaching in the field, ranging from the sophisticated to the simplified.

More sophisticated approaches, better suited for controlled research studies, generally comprise some form of the convection–dispersion equation (e.g., Rose et al., 1982):

$$\frac{\partial C}{\partial t} = \frac{D\,\partial^2 C}{\partial Z^2} - \frac{V\,\partial C}{\partial Z}$$

where C is the concentration of nitrate in the soil water, D is the dispersion coefficient, t is the time variable, Z is the soil depth, and V is the average velocity of pore water. In some cases, a nitrate plant uptake factor, B, may also be incorporated (Smith et al., 1984). Provided the necessary input factors are available, the sophisticated approaches can describe nitrate leaching quite well. For the uncalibrated field situation, however, the input factors are not always available, and simplified, more practical nitrate leaching approaches are required.

Simplified nitrate leaching approaches use the center of mass "piston displacement" technique, which has a wide experimental and conceptual basis (Davidson et al., 1978; DeSmedt and Wierenga, 1978). The principal assumption is that soil water above field capacity initially present in the profile is displaced ahead of the water entering the soil surface. Incorporation of the piston approach with pertinent soil and climatic factors allows for estimating nitrate leaching depth potentials on a general basis. The following example illustrates one simplified predictive method (Smith and Cassel, 1990). Basically, it involves incorporating a soil leaching potential with a climatic leaching potential.

The soil leaching potential reflects the water-holding properties of the soil and its capacity to permit infiltration. Figure 1 illustrates the climatic leaching potential. The data are for the Raleigh, North Carolina, area and represent mean

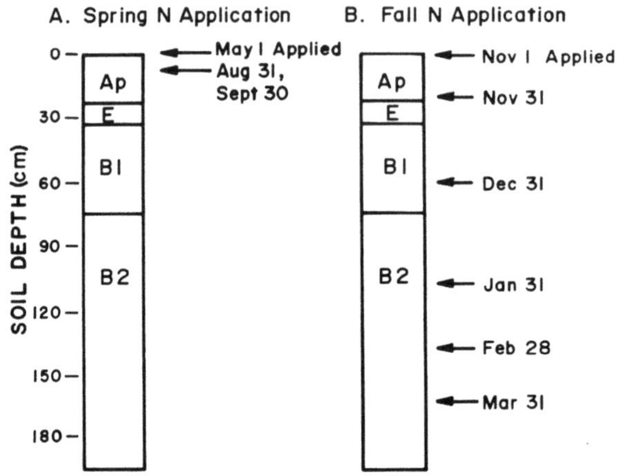

Figure 2. Probable depth of N fertilizer leaching for Norfolk soil at Raleigh, North Carolina (Smith and Cassel, 1990).

monthly precipitation (P) and potential evapotranspiration (PET). Note that the mean monthly PET exceeds P from late April to mid-October. Consequently, this is a period when, on average, water would not be available for leaching. On the other hand, P exceeds PET from November through March, thereby indicating good potential for leaching.

To estimate the probable depth of nitrate leaching for a particular location, only the soil water-holding properties, the associated P and PET, and the relevant management factors are required. Then, a simple water balance format (Smith and Cassel, 1990) is constructed. Figure 2 shows the probable depth of nitrate leaching using the format for nitrate fertilizer applied by surface broadcast to a Norfolk sandy loam at corn planting time or the preceding fall. Obviously, the probable depth of leaching is much greater (approximately 180 cm) when the fertilizer was applied in the fall. Moreover, by spring planting time the fertilizer N is already at a depth below the expected corn rooting depth, and would have good potential for leaching to the groundwater. With the spring application, however, the probable depth of leaching is <30 cm even at harvest time. Provided excess spring fertilizer N has not been applied, little nitrate would remain for leaching at this depth.

In the preceding example, no allowance was made for preferential leaching down soil macropores or for nitrate exclusion in the soil (Smith, 1972). Both phenomena may need to be accounted for with certain soils and conditions. However, the macropores can be particularly difficult to account for, because their formation and continuity depend on various factors, such as earthworms, plant roots, and soil drying. Likewise, stony, fissured, and karst topography soils (i.e., sinkholes at or near the surface) may provide direct flow channels for rapid movement of water and associated nitrate. In the case of irrigation, it is simply added in as a precipitation contribution. If desired, any soil-derived nitrate contributions may also be factored in. Situations involving deep nitrate leaching in the vadose zone are considered by Pionke and Lowrance (1990) and Shaffer et al. (1990).

D. Tile Drainage

Millions of hectares of poorly drained U.S. soils have been converted to highly productive croplands by tile drainage. Letey et al. (1977) reviewed nitrate-N concentrations in tile effluents from 55 sites in six areas of California. Average site concentrations ranged from 1 to 196 mg L^{-1}, with only about one-quarter of the sites averaging <10 mg L^{-1}. A review of N losses in tile effluents from the corn belt states of Iowa, Minnesota, and Ohio was conducted recently by Logan et al. (1980). They found annual nitrate-N losses were generally <30 kg ha^{-1}, but increased with N fertilizer applications above crop needs. Nitrate-N concentrations were above 10 mg L^{-1} even when 20 kg of fertilizer N ha^{-1} or less was applied annually to continuous corn (*Zea mays*). On the other hand, nitrate losses and concentrations with alfalfa (*Medicago sativa*) were very low. Therefore, periodic inclusion of alfalfa on the lands in continuous corn may be beneficial in maintaining nitrate losses in the tile effluents at more desirable levels. Little is known, however, regarding N losses when alfalfa is destroyed and the land is planted to other crops.

Because tile line effluent represents only a portion of the infiltrating water, the question may be raised regarding the fate of nitrate in the water bypassing the tile and continuing downward. Provided an organic energy source is present, the potential for denitrification exists, as observed by Gambrell et al. (1975) for a North Carolina coastal plain soil. Such processes do not always occur, however, because Hallberg et al. (1986) observed in Iowa that some high-nitrate tile concentrations reflected water table concentrations and that of infiltrating water which recharged the aquifer.

E. Volatilization of Gaseous N

Gaseous N losses involve complex processes associated mainly with ammonia volatilization and denitrification. The former is more prevalent with surface-applied N and the latter with incorporated N. Significant advances in gaseous N methodologies have been developed in the past 15 years. Techniques are now available for evaluating the magnitude of ammonia, dinitrogen, nitrous oxide, and nitric oxide losses as influenced by various cropping practices. Seldom, however, are all of the gases measured at one time for any major cropping system or cultural practice. A serious problem in making assessments is the spatial variability associated with soil heterogeneity. This brief review cannot cover in detail the many gaseous N developments and their application. Additional information can be found in publications by Bock and Kissel (1988), Freney and associates (1983, 1988a, 1988b), Mosier and Heinemeyer (1985), and Terman (1979).

1. Ammonia

Freney et al. (1983) have reviewed the chemistry of ammonia and the soil factors that affect ammonia volatilization, retention, and transport. Ammonia is highly soluble, very reactive, and its volatilization increases with increasing pH. The various reactions that govern ammonia loss (Freney et al., 1983; Vlek et al., 1981) can be represented as

$$\text{adsorbed } NH_4 = \text{solution } NH_4^+ = \text{solution } NH_3 = \text{atmospheric gaseous } NH_3$$

Besides pH, equilibria among the various species are effected by temperature, partial pressure of NH_3 in atmosphere, and NH_4^+ concentration in solution. The effect of pH on the equilibrium between NH_4^+ and NH_3 can be calculated from the equation

$$pK_a - pH = \log_{10} \frac{[NH_4^+]}{[NH_3]}$$

Emerson et al. (1975) provide an equation for calculating pK_a at a particular temperature:

$$\log_{10} K_a = -0.09018 - \frac{2729.92}{T}$$

where T is the absolute temperature (°K). The pK_a at 25°C is 9.246. Freney et al. (1983) indicated the relative percentages of NH_3 at pH values of 6, 7, 8, and 9 are approximately 0.1, 1, 10, and 50, and thus as the solution pH increases, greater potential for NH_3 volatilization exists. Temperature also affects the solubility of ammonia in water and the diffusion rates in soil. As temperature increases, diffusion rates increase, whereas ammonia solubility decreases. Its volatilization depends on the same environmental conditions as water evaporation. In fact, various investigators have shown that ammonia loss is directly related to water loss. Therefore, experimental setups to measure ammonia volatilization must not alter temperature, relative humidity, wind speed or similar factors.

Enclosures of various types have been used to measure ammonia loss in the laboratory and growth chamber with some success; however, field enclosures are considered unreliable because they interfere with wind speed, temperature, and water evaporation. It is known that ammonia is rapidly absorbed by plant leaves, so the size and development of the crop canopy influences atmospheric ammonia profiles. Also, NH_3 is rapidly adsorbed to glass, rubber, plastic, and other surfaces. Hence, tubing and various materials used for enclosures are likely to retain the molecule. Many investigators have used open acid traps, acid saturated sponges, or filter papers, but such techniques interfere with air movement and create artificial diffusion gradients, making it nearly impossible to properly determine ammonia emissions.

At present, micrometerological techniques that do not disturb the natural environment and integrate the ammonia flux over known areas are being used successfully to measure field emission of ammonia from upland and flooded soils. Denmead (1983) describes the micrometeorological techniques of gradient diffusion and mass balance. If there is a uniform source of NH_3 at the ground surface over a large area, a concentration gradient is developed in the atmosphere above the area. And if the conditions are constant with time, the upward transport of NH_3 can be calculated from measurements of the gradient diffusion by the equation

$$F = K\frac{dc}{dz}$$

where F is the flux density of NH_3, K is the diffusivity for NH_3, c is the atmosphere NH_3 concentration, and z is the height above ground where NH_3 is being measured. The equation is valid only when the flux is constant with height in the air layers above ground surface. This requires that the experimental area be large and uniform. The depth of the air layer in which the flux is constant with height is roughly between 1/100 to 1/200 of the fetch or distance upwind, which limits applications of the equation to fields of several hectares. The diffusivity, K, of NH_3 in air is the eddy diffusivity (movement of parcels of air from one level to another) and is several orders of magnitude higher than molecular diffusivity. Also, because its magnitude varies with environmental conditions, K must be measured in place (Denmead, 1983).

Mass balance is probably the simplest and most convenient of the micrometeorological methods, and is suitable for small areas. It is used where the flux from a treated area is expected to be quite different from a similar-sized nontreated area. The horizontal flux of NH_3 across a vertical plane of unit width on the downwind edge of a treated area is equated with the flux from a similar area upwind. The horizontal flux at any height is the product of horizontal wind speed and NH_3 concentration. The product of wind speed and NH_3 is modified by NH_3 emission (Denmead, 1983). The flux density of NH_3 can be expressed in the integral equation

$$F = 1/x \int_0^z uc\, dz$$

where F is the flux density, x is the distance traveled by the wind over the treated area, z is the height of the layer in the atmosphere, u is the wind speed, and c is the NH_3 concentration in excess of nontreated or background. To evaluate the integral equation, the profiles of u and c must be precisely defined at a number of heights. However, if the wind has a log or power law profile, it is possible to predict the flux from measurement of the NH_3 at one height, called ZINST. Denmead (1983) discusses ZINST in detail and the experimental evidence for its existence. Freney et al. (1985) have used simplified versions of the mass balance approach to measure ammonium volatilization from flooded rice (Oryza sativa) at ZINST (0.8 m), but the method has not been tested extensively for upland crops. They have also developed a bulk aerodynamic method to measure ammonia emission from flooded rice, but found the method was not as accurate compared to mass balance.

As noted earlier, NH_3 losses increase with increasing alkalinity or when ammonium salts produce microenvironments of high pH. In the dissociation of ammonium to ammonia,

$$NH_4^+ = NH_3 + H^+$$

a proton is produced, thus, as ammonia is lost from the system, the floodwater or soil media should become more acidic. The rate of acidification depends on the buffering capacity (base saturation and calcium carbonate content) of the medium (Avnimelech and Laher, 1977). Therefore, one can expect increased ammonia volatilization from calcareous soils that have a high base saturation, high carbonate content, or other forms of alkalinity. A number of carbonate reactions are important to ammonia volatilization (Terman, 1979; Fenn and Kissel, 1973; Fenn et al. 1981; Stumpe, 1981):

$$(NH_4)_2X + CaCO_3 = (NH_4)_2CO_3 + CaX$$

$$(NH_2)_2CO + 2H_2O + \text{urease} = (NH_4)_2CO_3$$

$$(NH_4)_2CO_3 = 2NH_3 + CO_2 + H_2O$$

Knowledge concerning the partial pressure of NH_3 and CO_2 that occurs upon the dissolution of ammonium carbonate or the hydrolysis of urea is necessary to predict or understand ammonia losses from flooded and upland soil. For example, the photosynthesis and metabolism of algae in floodwater can drastically alter the CO_2 partial pressure with diurnal cycles occurring because of the cycling of solar radiation. Also, the anion (X) associated with ammonium fertilizer influences the solubility of CaX. For example, if X is SO_4 or HPO_4 then CaX precipitates, driving the reaction toward ammonium carbonate, whereas, if X is NO_3 or Cl, the solubility of CaX is increased and the formation of ammonium carbonate is reduced, thus decreasing the volatilization of ammonium.

Knowledge of this information allows one to design a number of fertilizer practices that should reduce ammonia losses with ammonium salts, urea, and anhydrous ammonia. Ammonia volatilization can be slowed by restricting transport within the soil and evaporation at the soil surface. In both upland and flooded conditions research has shown ammonia losses can be reduced by placing ammonia-based fertilizers below the soil surface or, if the fertilizer is broadcast on the soil surface, immediately incorporating it by cultivation. If ammonium salts or urea are to be applied in irrigation systems times of high wind and high temperature, should be avoided—that is, fertigation should be applied during periods of low water evaporation, in the evenings or at night when the temperatures and wind speeds are at the minimums. Enough water (at least 2.5 cm) should be applied to move urea down into the soil. On calcareous soils ammonium salts, such as ammonium sulfate, that react to form ammonium carbonate and insoluble precipitates with calcium should be avoided. Each crop has a nitrogen requirement that varies as the crop matures. Applications of nitrogen timed closely to the crop requirement are most efficient. Usually, well-established roots are more effective at reducing inorganic N sources in the soil profile. In addition, foliar absorption of ammonium in the crop canopy can further reduce ammonia losses (Denmead et al., 1976). For rice plants recovery of applied urea is known to be much greater if the urea is applied at the time of maximum tillering or initiation of panicle (Heenan and Bacon, 1987; De Datta, 1981).

2. Nitrogen Oxides/Dinitrogen

Transfers of nitrogen oxides (NO, NO_2, N_2O) from the plant–soil system have gained increasing attention in recent years. This is because NO and NO_2 have been implicated in acid rain formation and the generation of photochemical smog, and N_2O has been implicated in the photochemical destruction of ozone in the stratosphere. The nitrogen oxides plus dinitrogen (N_2) are important nitrification and denitrification gaseous reaction products that decrease the efficiency of nitrogen use in crop production.

Various methods (Lemon, 1978; Denmead, 1979; Rolston et al., 1976; Ryden et al., 1978; Galbally and Roy, 1978; Galbally et al., 1987; Siegal et al., 1982; Mulvaney, 1984; Mulvaney and Boast, 1986; and Blackmer et al., 1982) have been used for measuring emissions for nitric oxide (NO), nitrogen dioxide (NO_2), nitrous oxide (N_2O), and dinitrogen (N_2). Flux measurements using the meteorological techniques outlined earlier for ammonia have not been very successful for nitrogen oxides and dinitrogen because analytical methods are not sufficiently sensitive to measure relatively small fluxes from the soil surface directly. The atmospheric reservoir of dinitrogen is extremely large and results in a tremendous dilution effect for $^{28}N_2$ and $^{29}N_2$. The atmospheric reservoir for N_2O is approximately 300 ppb, and is still large enough to cause considerable dilution of small fluxes of N_2O. Nitric oxide is readily oxidized, so its lifetime in the atmosphere may be in the order of seconds to minutes.

To overcome the analytical difficulties, scientists have turned to chambers of various designs to enhance atmospheric concentrations (Denmead, 1979; Ryden et al., 1978). Hutchinson and Mosier (1981) vented an enclosed cover to the atmosphere that exerted atmospheric pressure fluctuations on the soil, which may result in a pumping action on soil gases. As the nitrogen oxides or dinitrogen accumulate beneath the cover, the gas exchange between the soil and enclosed atmosphere should decrease. The researchers modified the flux equation to overcome this problem by measuring the change in concentration of the gaseous component with time. The problem can be further resolved by using short collection periods. However, short collection periods must then account for diurnal variations (Galbally et al., 1987; and Blackmer et al., 1982) that can be related to temperature changes in upland and flooded crops and the diurnal algal growth patterns, caused by diurnal light patterns in flooded rice. Furthermore, since fluxes are measured by enclosing very small areas of soil, small errors associated with the measurements may be magnified as data are extrapolated to large land areas. Analysis variability is caused by such soil factors as carbon content, pH, temperature, nitrate content, water-filled pore space, soil texture, and microbial ecology, which affect nitrification and denitrification. The enclosure volumes are also typically quite small and cannot accommodate large plants. Therefore, plant influences on loss patterns are usually not measured. These patterns may change as plants assimilate N, absorb water, and shade the soil. In the rice crop system, the plant may be a major conduit for gaseous N transport through the flooded system. In addition, little is known about how enclosures exclude short-term pressure fluctuations associated with turbulent winds which might affect gaseous fluxes.

Nitric oxide and NO_2 concentrations in an incoming airstream can be determined by chemiluminescent detection. Reactions are chemiluminescent when the excess energy contained in the products can be released in the form of light. The reaction between NO and O_3 produces NO_2 in an excited state, which relaxes to produce red light that can then be detected by a photomultiplier, permitting an NO determination in the order of 1 to 2 ppb. Nitrogen dioxide in the airstream can be detected by reducing it with a molybdenum reaction at 400°C to NO and then using a difference method to separate NO from NO_2. When nitrogen dioxide encounters a surface wetted with luminol (5-amino-2,3-dihydro-1,4-phthalazinedione) a chemiluminescence is produced, the luminescence coming from the luminol. In this case, NO_2 can be measured directly at concentrations in the 10 to 20 pptv. Using the luminol method, NO must be oxidized to NO_2 before detection and the two compounds are separated by difference.

From laboratory measurements it is clear that nitrification is a major contributor of NO. Hutchinson et al. (1987), in incubation studies, found that the addition of ammonia enhanced NO emission under aerobic but not under anaerobic conditions. The fraction of added ammonia lost as NO exceeded N_2O and was large enough to have important agronomic and environmental implications. Tortoso (1988), using inhibitors, has shown that NO is produced during the oxidation of ammonium to nitrate and emission rates increased as nitrite concentrations increased. When present in the soil, nitrite can enter into enzymatic and nonenzymatic reactions, leading to NO, NO_2, N_2O, and N_2 production (Cady and Bartholomew, 1960; Clark, 1962; Clark and Beard, 1960; Clark et al. 1960; Chalk and Smith, 1983; Wijler and Delwiche, 1954; Nommik, 1956; Smith and Clark, 1960; Stevenson and Swaby, 1964; Stevenson et al. 1970; Nelson and Bremner, 1969; Porter, 1969). Blackmer and Cerrato (1986) treated 28 autoclaved soils with nitrite and found amounts of NO ranging from 1 to 35 mg kg^{-1}, statistical analyses indicating that 95% of the variability for NO production could be explained by considering only soil organic matter content, pH, and their interactions.

During the past decade few field measurements of NO and NO_2 have been made, and the limited measurements make it difficult to judge their importance in the N cycle. Galbally et al. (1987) studied the emission of NO from fertilized rice and found only 0.002% of the fertilizer N was lost during the 10-day study period. They observed that the NO_x flux was very low for the first few days after application of urea and about five times higher during the period when significant nitrite and nitrate were present in the floodwater. Tortoso et al. (1986), under well-aerated conditions in nonfertilized systems, found that NO production followed the order of coniferous forest > native shortgrass prairie > fallow agricultural fields. They also noted that natural rather than agricultural ecosystems have a greater potential for aerobic loss of N oxides. They observed that NO emission rather than N_2O is the principal gas lost during nitrification and that the loss of NO during microbial denitrification may be higher than is commonly believed. Hutchinson and Follett (1986) studied the effects of long-term cultivation practices on NO emissions and found the emission rates followed the order of native sod > no-till > stubble mulch > conventional plow treatments. Again, NO emissions were greater than N_2O emissions and were closely related to nitrification rates. Anderson and Poth (1987) found that disturbed ecosystems (burned and unburned chaparral) that display high rates of mineralization and nitrification exhibit losses of NO and N_2O. Blockage of nitrification by acetylene again demonstrated that emissions of NO and N_2O result from nitrification.

Nitric oxide and NO_2 can be formed from nitrification reactions, chemodenitrification or self-decomposition of nitrous acid and denitrification (Smith and Chalk, 1980; and Nelson and Bremner 1970). The oxidation of NO and NO_2 and hydration of NO_2 during the diffusion of these gases through the soil results in some nitrate formation (Nelson and Bremner, 1970). Field measurement flux rates for NO and NO_2 (Slemr and Seiler, 1984; and Delany et al., 1986) were found to be strongly dependent on soil surface temperatures, with minimums in the morning and maximums in the early afternoon. These investigators have shown that the soil can be both a source and a sink for these gaseous N oxides.

Gas chromatographic analysis with an electron capture detector (ECD) or an ultrasonic detector, a carrier gas of 95% argon–5% methane, and Q porapak columns is a sensitive and precise method for determining N_2O (Rassmussen et al., 1976; Blackmer and Bremner, 1977). Mosier and Mack (1980) used 10 port-switching valves and two porapak columns so only N_2O is passed through the ECD, thus preventing oxygen interfaces with the detector. They also backflush their column to remove Freon, which after a few samples begins to leak from the column and causes background drifting. Gas samples containing 400 ppb N_2O can be determined with a precision of 3 ppb in 4 to 5 minutes. Space limitations do not allow us to discuss the numerous other methods that have been used to estimate N_2O fluxes; the reader is referred to Ryden and Rolston (1983) and Mosier and Heinemeyer (1985), who describe the methods and their advantages and disadvantages. The closed cover method is used to obtain flux rates of N_2O, but it suffers from the shortcomings mentioned earlier for NO.

Laboratory and field studies have provided extensive information (Sahrawat and Keeney, 1986) on factors that affect evolution of N_2O, and a number of these factors were enumerated previously. It is probably sufficient to indicate that there are three major sources of nitrous oxide: (1) the interactions of nitrite with soil organic matter and pH (chemodenitrification); (2) the production of nitrous oxide during the oxidation or nitrification of ammonium under aerobic conditions; and (3) the production of N_2O as nitrate is biologically reduced under the conditions of >70% soil water-filled pore space (Linn and Doran, 1984), low oxygen concentration, and a soluble carbon source to provide energy for the reaction. Since the emission of nitrous oxide may be derived from three totally different reactions, it is difficult or essentially impossible to extrapolate laboratory measurements to field conditions under which these three pathways could operate intermittently under variable soil conditions and substrate concentrations. Plants may further complicate the picture, as they affect the pools of available nitrate and ammonium, alter the soil moisture content, excrete soluble carbon compounds, and affect the oxygen status of the soil. For example, it is known that in flooded rice fields the hollow stem of the rice plant is a conduit for oxygen movement to the roots and N_2O and N_2 from the roots to the atmosphere. The rhizosphere zone around rice roots may be a mosaic of oxidized and reduced areas. In the field, when soils are warm and mineralization and fertilization provide ammonium substrate, then NO and N_2O production proceed at a low steady rate as long as conditions are favorable to nitrification. This pattern may abruptly change to one of denitrification in a few hours following irrigation or rainfall. Sharp fluxes from the soil surface can be observed within 24 to 48 hours, but the peak rate depends on the ease with which various gases are able to diffuse to the soil surface, and it slowly declines over days as aerobic conditions begin to return. Rapid uptake of nitrates by plants can limit the substrate for denitrification. Banded urea, anhydrous ammonia, and ammonium sulfate or phosphate can create zones where pH varies widely. In alkaline and poorly buffered soils the pH may increase drastically with increasing concentrations of these compounds. Nitrite oxidizers (*Nitrobacter*) growth is suppressed at higher pH, apparently due to ammonia toxicity, while growth of ammonium oxidizers (*Nitrosomonas*) is not affected. This results in accumulations of nitrite, with its possible chemodenitrification reactions (Chalk and Smith, 1983).

Nitrogen use efficiency and denitrification studies require direct in-field estimates of dinitrogen and NO_x. Total gaseous emissions can be measured with the aid of highly enriched N-15 salts or urea. The idea for the technique was first proposed by Hauck et al. (1958) and further expanded by Hauck and Bouldin (1961). The method involves applying highly enriched 20 to 99 atom % N-15 salts and urea to soil. If nitrification and denitrification products from the fertilizer give rise to gaseous N, it will be selectively enriched, since the biological or chemodenitrification products are not randomly mixed with atmospheric dinitrogen or with the organic N of the soil. Using this nonrandom N-15 distribution method permits calculations (Siegel et al., 1982; Mulvaney and Kurtz, 1984; and Mulvaney, 1984) of the N-15 enriched gas from fertilizer as well as gas from nonlabeled soil N constituents. The gaseous products may all be reduced to dinitrogen, or they may be selectively separated by freezing techniques and measured separately.

Field measurements have shown large variability associated with NO, N_2O, and N_2. There are seasonal patterns associated with rainfall or irrigation and plant growth variations, as well as substrate production from organic matter mineralization, and these are compounded by the timing and placement of fertilizer application. As yet, there is considerable uncertainty about the seasonal or yearly emissions of dinitrogen and nitrogen oxides. However, general information indicates annual cropland losses for the two major gases, N_2 and N_2O, are in the range of 5 to 25 and 0.1 to 3 kg N ha^{-1}, respectively. For N_2/N_2O evolution during denitrification, therefore, we ascribe no general ratio. In the case of N_2O, it should be noted that some may be lost from soil by dissolution in drainage water, perhaps in amounts comparable to that lost in the gas phase at the soil surface (Dowdell et al., 1979).

IV. Practices to Ameliorate N Losses or Increase N Use Efficiency

A. Crop N Requirements

Crop N uptake determinations provide the basis for estimating N requirements which can then be adjusted to compensate for other N sources. The crop N requirement factor can be defined as crop N uptake near maximum yield (Meisinger, 1984). For corn (*Zea mays*) and wheat (*Triticum aestivum*), producers frequently use respective factors of 1.25 and 2.22 lb.N bu^{-1} (22 and 37 kg N Mg^{-1} grain). The values vary somewhat by region, and may require site-specific modifications to obtain economically and environmentally acceptable yields. When possible, yield goals should be based on prior history of the soil. A realistic yield goal is to use the average production over a 5-year period, and add no more than 5% to that production (Wiese et al., 1987). Additional details regarding impacts of cultural practices, methods of fertilizer application, crop rotations, and other considerations on crop development and N use efficiency have been compiled by Hauck (1984). In the future technological advances and more sophisticated management practices will probably be required to address anticipated environmental concerns.

Efforts to improve N efficiency will require better synchronization between soil N availability and crop N requirements. Therefore, cultural practices, soil water status, crop pests, and many other factors that affect crop N uptake patterns will complicate management decisions. Genetic selection for improved N efficiency in crops such as corn and sorghum (*Sorghum bicolor*) may reduce N requirements, but efforts to quantify N $\times$ genotype interactions can easily be dominated by environmental interactions (Duvick and Cavalieri, 1987). Efforts to modify plant N metabolic processes by manipulating genetic material in a way to improve N efficiency may be several decades away. Until then, computer simulation of crop growth, N uptake, soil biological and chemical processes, and water status provides an alternative way to evaluate management practices that impact on nitrate leaching and groundwater quality (McIsaac et al., 1984).

B. Soil N Availability Tests

Available soil N represents residual nitrate in the profile, plus N mineralized from the soil organic matter during the growing season. While residual nitrate has proven to be a useful index in dryland regions (i.e., Plains states and Western United States), no generally accepted index exists for N mineralization. Obviously, such development would represent a major advance for avoidance of excessive fertilizer N application.

A complement to a soil N test may be a plant tissue N test. An attractive feature of tissue tests (Rauschkolb et al., 1984) is that the plant root system tends to integrate spatial variability of soil N supplying power over a relatively large field volume. To date, tissue tests have found application mainly as a diagnostic tool to indicate N sufficiency/deficiency for high-value, specialty crops, rather than field crops. Nevertheless, N tissue tests for cotton (*Gossypium hirsutum*) (Tucker and Murdock, 1984) and corn (McClenahan and Killorn, 1988; Schepers et al., 1988) appear to be receiving increasing acceptance.

1. Residual Nitrate

Tests for residual nitrate are useful for soils suspected of having an appreciable accumulation of nitrate in the profile. These tests measure only that quantity of nitrate present at the time of sampling. Consequently, they have limited utility as a soil N index under conditions conducive to leaching. From a Midwest soil test standpoint (Whitney, 1970), profile nitrate should be sampled in row crop fields generally between November 1 and May 1. Winter small grains can be sampled after August 1 for preplant N application and after November 1 for top dressed N. Spring small grain fields would be sampled after October 1 until freeze-up, or in spring after the soil thaws. Additional considerations include sampling after preirrigation, and sampling sandy soils in the spring for row crops and top dressing of wheat. In all cases, the subsoil as well as the surface soil is sampled. Some laboratories will not conduct the nitrate soil test unless a minimum depth of 60 cm has been sampled. Also, depending on the laboratory, more detailed information about the sampled area may be requested.

In the case of side dressing only, surface soil samples (30 cm depth) collected after corn planting are being evaluated as a means of determining the amount of N application (Magdoff et al. 1984). Also, soil nitrate determination and N fertilizer application on-the-go have been reported for corn production in Illinois (Anonymous, 1989). An electronic sensor mounted on a moving soil shank measures nitrate directly and transmits a signal to a microcomputer that controls the fertilizer N application.

2. Soil Organic N Availability

A significant portion of a crop's N needs is supplied by mineralization of soil organic matter during the growing season. As noted earlier, this mineralization usually constitutes less than 3% of the total soil organic N. An elusive goal has been to conveniently determine the amount of potentially available soil organic N, and use it as a base for a more precise fertilizer N recommendation. Various N availability indexes exist (Keeney, 1982b), but they typically provide a qualitative rather than a quantitative measure of soil organic N availability. In the early

1970s, Stanford and Smith (1972) approached the problem using a soil N mineralization potential (N_0) concept. Their intent was to determine a soil's N_0 under optimum conditions, and then adjust it for field moisture and temperature. In this manner, they hoped to be able to estimate how much soil organic N would become available during a prescribed growing season. Amounts of N mineralized for 39 agriculturally important U.S. soils were determined using a successive aerobic incubation procedure for 30 weeks. The N_0 value for each soil was then estimated on the assumption that N mineralization obeyed first-order kinetics, and the accepted value of N_0 was that which resulted in the best fit for the linear relation between log (N_0–N_t) versus time. For surface soils, N_0 ranged from 20 to more than 300 mg kg^{-1}.

Subsequently, considerable attention was devoted to determining short-term mineralization procedures for N_0, chemical extraction procedures for N_0, using the concept to model N cycling in soils, and testing its application. Overall, application tests have been mixed, but with some success in situ (Herlihy, 1979), the greenhouse (Stanford et al., 1973), and the field (Oyanedel and Rodriguez, 1977; Smith et al., 1977). More recent tests indicate use of undisturbed soil in the mineralization procedure provides an N_0 better suited to field conditions (Cabrera and Kissel, 1988). Also, a rapid steam distillation, borate buffer method of assessing potentially available organic N has been introduced (Gianello and Bremner, 1988) that is not significantly affected by drying of the soil.

The original concept of N_0 has also been modified by some workers (e.g., Molina et al., 1980) to include more than one soil organic N pool. However, recent work in Canada (Campbell et al., 1988) indicates a one-component N_0 model is satisfactory for quantifying N fertilizer requirements. At present, also, attention is being given to an electroultrafiltration (EUF) approach to estimate N_0 (Schepers and Saint-Fort, 1988). Nevertheless, to date, no soil organic N availability procedure has received general acceptance from a soil test standpoint. Ultimately, a systems type, mass balance N approach (Meisinger, 1984) may prove to be the best alternative. Presently, the recommendation is to follow pertinent state N fertilizer guides, which are geared to specific crop needs and soil areas.

3. Sampling Variability

The importance of obtaining a representative sample for soil N characterization cannot be overemphasized. State recommendations for soil test purposes vary, but generally range from 10 to 20 composite surface samples per soil area sampled. For deeper profile samples, fewer may suffice because soil N variation and content tend to decrease with increasing profile depth. An aerial photo of the field in question is a good aid in delineating uniform areas for sampling purposes. These photos, available through SCS or ASCS offices on a scale of 1:7920 (8 in. per mile), have been taken during the dormant season. They often provide a clear delineation of the soil series, slope, drainage pattern, and land use. Such information, particularly when incorporated with Geographic Information Systems

(GIS) (Peterson, 1989), can be of considerable value in locating uniform field areas for sampling purposes.

Even small, relatively uniform soil areas can be quite variable, with nitrate coefficients of variation in the 30 to 50% range not uncommon (Keeney, 1986a). More sophisticated, geostatistical techniques for ascertaining the number of soil samples required for a given precision and accuracy may use the samples' spatial dependence (Nielsen et al., 1982). Such techniques involve the use of autocorrelograms and the concept of a regionalized variable. With the geostatistical approach, an estimate for the N characteristic at an unsampled site is determined from measurements made nearby, rather than from a site average. Application of geostatistics to soil science is still relatively new, and the results to date have often given answers similar to those obtained from more conventional techniques (Warrick et al., 1986). A major drawback is the cost, because of the greater number of samples required in the analysis.

C. Nitrogen Fertilizers

A wide variety of N fertilizer formulations is available to producers to accommodate various times, rates, and methods of application. Anhydrous ammonia (gaseous) accounts for approximately 40% of all fertilizer N marketed in the United States (Hargett et al., 1988). This value increases to over 60% in Kansas, Nebraska, Illinois, and Iowa, largely because anhydrous ammonia is the least costly form of fertilizer N. Granular N fertilizers such as urea and ammonium nitrate are easier to store and handle, but more costly. Liquid N fertilizers (e.g., urea ammonium nitrate, starter blends) are frequently the most expensive but more versatile in terms of blending with other fertilizers or herbicides, plus they can be easily injected into sprinkler irrigation systems. Fertilizer N formulations and methods of application are discussed by Tisdale et al. (1985); however, special fertilizer N management considerations may be required to minimize contamination of groundwater. Management systems that hold promise for the future include the use of satellite imagery [e.g., Global Positioning System (GPS)] and grid farming, which allow applying N inputs by soil variations rather than by fields (Reichenberger, 1989; Reichenberger and Russnogle, 1989).

The foremost consideration when developing N management practices to minimize nitrate leaching must be to determine when water is likely to percolate below the root zone. Irrigation is frequently implicated as the cause of nitrate leaching; however, precipitation between April 1 and June 15 in much of the northern plains exceeds ET (Schepers, 1988). Therefore, the benefits of fertilizer N applications before or during this vulnerable period for percolation must be evaluated in terms of crop N requirements and risk of leaching. Nitrate leaching can be reduced when a nitrification inhibitor (i.e., nitrapyrin, DCD) is added to ammonium fertilizers to slow the rate of biological conversion to nitrate (Keeney, 1986b; Owens, 1987). Effects of nitrapyrin on corn yield are frequently positive when nitrate leaching losses would otherwise be large. In contrast, if early season leaching losses are small and the crop was not N deficient because of adequate or excess N availability, then crop response to nitrification inhibitors may be minimal (Hergert and Wiese, 1978; Touchton and Boswell, 1978; Nelson and Huber, 1978; Papendick and Engibous, 1978; Onken, 1978).

Starter fertilizers frequently promote enhanced early season growth, but they may not affect yield because the amounts of N in starter fertilizers placed near the seed are only a small part of the total crop N requirement. Crop utilization is usually the greatest with side-dress fertilizers because application time and crop uptake are closely synchronized. Although side-dress N application may be the preferred method of fertilization, split applications or starter fertilizer may be required to meet early season crop N requirements. Somewhat greater broadcast N application rates are frequently required when crop residues have the potential to immobilize N, as under no tillage conditions (Doran and Smith, 1987). For this reason, greater fertilizer N uptake by crops can be achieved with surface band applications or with subsurface knife applications. Crop clearance by tractors usually restricts fertilizer applications requiring large amounts of power, however, high clearance vehicles make it possible to apply some forms of fertilizer virtually any time. Wheel-type spoke injector mechanisms mounted on elevated tractors or similar vehicles permit subsurface placement of liquid fertilizer at virtually any stage of crop growth (Baker et al., 1985). Foliar application of N fertilizer to row crops and small grains has had limited effects on yield. Below et al. (1984a, 1984b) found that tagged-N foliar fertilizer applied to corn at anthesis was recovered in the grain, but yield and N concentration were not affected by the foliar N application. In this case, the crop was not N deficient and the foliar absorbed N-15 substituted for other N in the crop.

Nitrogen fertilizer along with genetic improvement and irrigation are probably the three most significant factors responsible for yield increases over the past five decades (Carlone and Russell, 1987). Cultivar selection for crops requiring large amounts of N, such as corn, has traditionally been made under high-fertility conditions. Occurrence of nitrate in groundwater has prompted interest in more N efficient cultivars. Corn yields from most N × genotype studies are not conclusive, largely because environmental interactions are difficult to quantify and frequently dictate the nature of the yield response (Duvick and Cavalieri, 1987). Therefore, research to select genetically superior cultivars for improved N efficiency is proceeding cautiously as the potential for improvements using genetic engineering and molecular biology techniques are evaluated. Both field (Olsen, 1986) and hydroponic studies (Below and Gentry, 1987) indicate the benefit of mixed ammonium and nitrate nutrition for corn. Metabolic reasons for the increased corn yield attributed to mixed N nutrition are uncertain (Hageman, 1978), but some cultivars respond by increasing kernel number, others by increasing kernel size, and some by a combination of the two (Schepers and Below, 1988). Nitrification inhibitors can be used to modify the proportion of nitrate to ammonium in soil on a temporary basis, but the proportion of these N forms may be adequate under natural conditions, depending on soil chemical and physical conditions.

Time of fertilizer application can have a major effect on yield and an implied effect on nitrate leaching. The best results for rain-fed winter wheat in Colorado were obtained by applying N early enough to encourage leaching so that a majority of the stored water and N were available at similar depths in the root zone (Smika and Grabouski, 1976). Under irrigation, greatest N utilization by corn is usually obtained with side-dress application, which corresponds to the time of greatest N uptake. Fertigation can also be timed to coincide with the time of greatest crop N requirement, but may result in volatilization of ammonia sources (Denmead et al., 1982) and the convenience of fertigation may promote N application beyond the time of greatest utilization. As noted earlier, crop N status during the growing season can be evaluated by tissue-testing procedures. Such procedures are routinely used for some fruit and nut crops, sugar beets (*Beta vulgaris*), and potatoes (*Solanum tuberosum*). Application of tissue-testing procedures to improve N management holds promise for other crops such as corn and sorghum (Schepers and Below, 1988) because leaf analysis for total N is an easy and rapid way to monitor crop N status, which is a reflection of N availability, regardless of the source.

Mineralized N is the most difficult source of N to quantify because of its dependence on tillage, climate, and substrate availability. Nevertheless, mineralized N can amount to a large part of crop N requirements on fertile soils, and if not used it can contribute to nitrate contamination of groundwater. The rather slow and perpetual nature of mineralization and immobilization requires special consideration to be synchronized with crop N requirements when large amounts of crop residue are added to soil. Residues with less than 1.2% N usually immobilize N (Broadbent, 1978), which reduces the potential for nitrate leaching, but they can also create a N deficiency before mineralization dominates. Green manure cover crops, legume vegetation, and animal manures usually contain adequate N to promote mineralization shortly after soil incorporation. The fertilizer N equivalence for legumes ranges from less than 20 kg N ha^{-1} for some interseeded legumes (Hesterman, 1988) to over 100 kg N ha^{-1} for several winter cover crops (Frye et al. 1988). Additional information regarding N use and crop residues can be found in Power and Doran (1984).

D. Nitrogen/Soil Conservation Practices

1. Conservation Tillage

Conservation or reduced-tillage continues to increase in usage as an alternative for practically all forms of crop production (Follett et al., 1987). Management systems that maintain crop residues at or near the soil surface have several attractive features, including less on-farm energy use, more available soil water, and reduced soil erosion. Concurrently, the adoption of conservation tillage practices has stimulated interest in associated N environmental impacts.

There is no question that conservation tillage is effective in reducing particulate N losses associated with runoff (Gilliam and Hoyt, 1987). Berg et al. (1988), for example, observed a 21- and 8-fold reduction in sediment and total N loss, respectively, on highly water-erodible wheat soils in the southern plains, with no-till compared to conventional-till (disking and sweeps). This occurred over a 2-year, above-average rainfall period. The N loss reduction was not as great as the sediment loss reduction because of humus enrichment, selective removal of fines, and less dilution by infertile subsoil. In the case of soluble N in the runoff, differences between the tillage treatments were not that great, 1.7 and 1.9 kg NO_3–N ha^{-1} yr^{-1}, for conventional and no-till, respectively. Such differences are generally small, provided the fertilizer N is incorporated into the soil (Gilliam and Hoyt, 1987).

Effects of conservation tillage on leachable N are not as well delineated as for surface runoff N. Generally, conservation tillage provides a wetter, cooler, more acidic, and less oxidative soil environment. Under such conditions, the processes of ammonification and denitrification may be favored over nitrification. On the other hand, for the nitrate present, the leaching potential may be greater with reduced tillage. This is because more undisturbed, large soil pores exist for nitrate movement. Dick et al. (1986), for instance, observed twice as much water flow down, out of the root zone in no-till than in conventional-tilled soils. This higher flow was attributed to reduced evaporation from the soil because of the surface residues and increased number of channels continuous with the soil surface. The surface mulch enhanced the environment for earthworms, and the lack of tillage preserved the flow-transmitting capacity of their burrows for several years.

The role of macropores regarding leaching under conservation tillage can be illustrated by two contrasting examples. In the first case, nitrate leaching was found to be less under no-till than under conventional-till conditions, when nitrate was distributed on the soil surface (Kanwar et al., 1985). Under no-till, much of the water flowed through macropores to the subsoil, leaving most of the remaining nitrate distributed throughout the soil surface. Under conventional-till, no large macropores or channels leading from the soil surface to the subsoil were present, so the infiltrating water moved down the soil profile as a front, carrying the nitrate with it. The net result was less nitrate leached under no-till. In the second case, nitrate leached deeper in a killed sod, no-till system than under conventional-till (Tyler and Thomas, 1977). In this case, soluble ammonium nitrate fertilizer had been applied to the soil surface. Rainfall dissolved the fertilizer, and the water flowed through root channels from the soil surface, carrying the nitrate deeper down the profile. For the two examples, then, we see that the nitrate leaching process can be very different for similar tillage systems, depending on whether or not the fertilizer finds its way to the macropores.

Equipment development and research are currently under way to precisely and accurately position N fertilizer under conservation tillage to reduce soluble N leaching (Baker, 1985). In the case of ridge-till, for example, there may be an advantage in placing the N fertilizer in the ridge, because a large part of the water infiltrates the valleys between the ridges. Another possibility may be placing the fertilizer under the soil surface, in a coarse-textured band (Bowers et al., 1975).

2. Rotations, Cover, and N Scavenging Crops

Rotations and cover crops, historically used as a means of conserving soil and/or providing an organic N source, have received renewed interest as an aid in avoiding excessive N losses to the environment. Monocultures of grain crops (e.g., corn and wheat) require high inputs of fertilizer N, but such inputs can be reduced with crop rotations, which require less or produce organic N (Papendick et al., 1987). Voss and Shrader (1984) note that first year, non–N fertilized corn in Iowa following alfalfa may outyield fertilized monoculture corn. Concurrently, because less excess profile N is expected with a rotation, there should be less potential for N leaching. An exception is under certain rotation–fallow conditions designed to conserve water in drier areas. The saline seeps of the northern Plains, for example, are a result of a rotation–fallow practice allowing deeper water penetration that carries native solutes, including any profile nitrate present, to a shallow impermeable layer. The solutes can then move to a downslope position, resulting in the seep (Olson, 1986). It should be noted that such seeps are now being observed with increasing frequency in the southern Plains (Naney et al., 1986), especially on terraced fields.

Winter cover crops can be effective in absorbing both nitrate and available water during the fall, winter, and spring, thereby reducing the N leaching potential. When the cover crop is returned to the soil, some of the absorbed N is then available to the following crop. Both legumes (e.g., vetches, clovers, peas) and nonlegumes (e.g., small grains) are used, but from a strictly N leaching standpoint, rye (*Secale cereale*) appears to have certain advantages. It is the most cold tolerant, with lodging and disease posing little concern.

While an annual crop, such as rye, can be effective in scavenging excess available N from within crop rooting zones, deep-rooted perennials should be considered for nitrate accumulations below normal rooting depths. Alfalfa (*Medicago sativa*), with a potential rooting depth in excess of 5 m, is a crop that merits particular attention (Mathers et al., 1975; Olsen et al., 1970; Schertz and Miller, 1972; Stewart et al., 1968). In fact, soil nitrate contents under alfalfa can be lower than those under native grasses (Table 1).

3. Filter Strips

Vegetative filter strips, also referred to as buffer strips and riparian zones, are defined as strips or areas of vegetation for removing sediment, organic matter, and other pollutants from runoff and wastewaters (SCS, 1984). Under field conditions, excess runoff from terraces is frequently diverted to a strip. When it enters the strip, both the flow velocity and transport capacity of the runoff are reduced. The sediment and its associated pollutants are then removed from the runoff by filtration, deposition, infiltration, sorption, decomposition, and volatilization processes. The size of a filter strip needed for a particular situation may be calculated (Ohlander, 1976), but 30 m is a common width.

The effectiveness of filter strips in removing sediment and particulate waste N is well established (Neibling and Alberts, 1979; Lowrance et al., 1984; Robinson

Table 1. Nitrate contents in soil profiles associated with landuse[a]

Land use	No. soil profiles	Nitrate-N in 6-m soil depth[b] (kg ha^{-1})
Alfalfa	13	88
Native grass	17	100
Cultivated drylands	21	290
Irrigated fields[c]	28	562
Feedlots/corrals	47	1594

[a] Study conducted in Colorado (Stewart et al., 1968; Viets, 1975).
[b] Values represent averages for the soil cores.
[c] Does not include irrigated fields in alfalfa.

and Brockway, 1980; Young et al., 1980). For example, Robinson and Brockway (1980) evaluated filter strips on 18 research plots and observed that a 2.4-m small-grain strip achieved sediment removal rates of 80+% from furrow-irrigated beans and reduced the mean particle diameter in the runoff sediment from 64 to 17 μm.

Results regarding the effectiveness of filter strips for removing soluble N forms in runoff are mixed. Omernik et al. (1981) question their effectiveness, pointing out that soluble nutrients associated with sediment may move to streams over time by means of subsurface flow. Lowrance et al. (1983, 1984) studied water-borne nutrient budgets for the riparian zone in an agricultural area. They found that the riparian zone acted as a filter for nitrate, which was transformed to organic form. Similar results have been reported in New Zealand (Howard-Williams et al., 1986) where more than 70% of the nitrate uptake over time was attributed to aquatic macrophytes in the riparian zone. Obviously, some denitrification may also be occurring. Another factor influencing the efficiency of a filter strip in removing soluble N forms is how fast they pass through the vegetation. Incorporation of an N-scavenging crop, such as alfalfa, in the filter strip may retard vertical movement of the soluble N.

4. Terraces, Settling Basins, and Impoundments

These structures also provide a means for trapping particulate and soluble N forms in runoff. Schuman et al. (1973), for example, observed that N losses from level-terraced corn were only one-tenth those from contour-planted corn. In the case of settling basins, Edwards et al. (1980, 1983) channeled runoff from a 243-m^2 beef feedlot through a settling basin and determine the basin-reduced particulates in outflow by 54%. For major storm events, the basin was more effective than filter strips. Tile drainage of small sediment basins built into terraces also allows the discharge water to be relatively low in particulates and nutrients (Schepers et al., 1985).

With large impoundments, such as lakes and reservoirs, the emphasis is more on maintenance of a high-quality water supply for potable, irrigation, and recreation purposes than on their use as a settling basin. Nevertheless, even these types of structures serve a major role in the settling and removal of N components. For example, the geometric mean of total N concentrations in waters of 87 European and North American lakes was determined to be about one-half that in inflows, with the reductions in concentrations generally being greater for longer residence times (OECD, 1982). Less information is available for smaller agricultural field impoundments, but Menzel et al. (1986) recently determined concentrations of nitrate, ammonium, and Kjeldahl N in runoff and two associated SCS impoundments. The impounded water and outflows had lower N concentrations, except for similar ammonium concentrations. Total N (Kjeldahl plus nitrate) concentrations in the outflows were about 20% of that in the inflows. Similarly, Olson et al. (1973) observed that nitrate N content of inflow to three Nebraska reservoirs averaged 1 mg L^{-1} whereas the nitrate N content of outflow was only 0.05 mg L^{-1}.

E. Irrigation

Approximately 17% of all cropland in the United States is irrigated (Olson, 1986). Irrigation is usually intended to provide supplemental water to the crop or to leach excess salts from the root zone. Although irrigation may reduce the risk of crop failure, it can subsequently increase the potential for nitrate leaching and risk of groundwater contamination. Unfortunately, soils most susceptible to leaching are often those that are irrigated. The most positive approach to controlling nitrate leaching is to optimize the leaching fraction. Where soil salinity is a factor, the leaching fraction should be just sufficient to maintain an appropriate salt balance. Both time and method of irrigation must be considered as an integral part of N management practices.

Irrigation methods that rely on gravity for distribution of water (e.g., furrow and border) typically result in temporal and spatial variability in application rates (Elliott and Walker, 1982), especially on nearly level fields where water spreading may be a problem. Soils with moderate to rapid infiltration rates accentuate water distribution problems and can lead to portions of the field that are under water stress while in other areas the crop can exhibit N deficiency symptoms because of nitrate leaching. Practices to improve water application uniformity of gravity systems include reducing the length of run, land grading to establish uniform slopes, and reducing surface roughness to increase the rate of water advance across the field (Kemper et al., 1982). Efforts to increase uniformity of water application with gravity irrigation systems are frequently coupled with runoff reuse systems. Recently, computer-operated diversion valves in gated pipe distribution systems (surge irrigation) have made it possible to apply water at high rates for short periods of time to rapidly wet the soil and thereby promote soil sealing (Miller et al., 1987). Subsequent water application can be timed to match the desired application rates with the soil water-holding capacity of the root zone to minimize both percolation and runoff.

Runoff water from furrow-irrigated land is usually only slightly enriched in nitrate as it passes across the field compared to nitrate concentrations in the water source. Similar trends are noted for nitrate enrichment of runoff originating from precipitation, while ammonium in any water source is usually sorbed as it passes across the landscape (Schepers and Fox, 1989). The exception, regardless of the water source, is runoff generated immediately after surface application of N fertilizer or manure (Schepers and Francis, 1982). Uncertainties regarding prediction of nutrient discharge in runoff arise because of differences in antecedent soil moisture conditions and the degree of mixing during the runoff process (Wallach et al., 1988).

Other methods of water application such as level basin and sprinkler irrigation are usually tailored to a specific terrain or type of crop. Sprinkler irrigation systems often provide the greatest flexibility in terms of time and rate of application, which should minimize the risk for nitrate leaching. Many high-pressure sprinkler systems which are costly to operate have been replaced by medium- and low-pressure systems to reduce energy costs; however, these systems lead to additional runoff or infiltration (Gilley and Mielke, 1980). The relatively uniform water application possible with sprinkler systems makes them ideal for application of N fertilizer (fertigation) or pesticides at a time when the crop is too large for traditional application methods. Properly timed N applications in small amounts of frequently applied water can reduce leaching losses (Rhoads and Stanley, 1981) and should place less risk of contamination on groundwater than preplant N applications. It is important to note that precipitation immediately following irrigation can result in excess soil water and subsequent leaching, regardless of how carefully irrigation is managed. For this reason, irrigation scheduling procedures frequently attempt to maintain a small storage reserve within the root zone to accommodate a limited amount of precipitation without causing leaching.

Drip or trickle irrigation is probably the most water- and nutrient-efficient method of irrigation, but it is also the most costly to install (Mitchell, 1981). It is therefore frequently used on high-value crops [e.g., orchards, vineyards, strawberries (*Fragaria chiloensis*)] for which installation cost can be amortized over several years, the quality of the saleable product can be reduced by sprinkler application, and water distribution and soil aeration are problems with gravity irrigation. Nutrient additions to soil through drip irrigation systems should minimize nitrate leaching provided excess water applications are not required to maintain a leaching fraction to regulate salt accumulation.

Special water management considerations are required near the end of the growing season when precipitation during the fallow period is likely to cause leaching. Therefore, irrigation scheduling procedures should attempt to end the growing season with as large a water deficit as possible without reducing yield. Coarse-textured soils, with their low water-holding capacities, are especially difficult to manage with respect to nutrients and water because of the frequently shallow root systems. For these reasons, rate and time of water application must be determined for each soil, method of irrigation, and type of crop. Fertigation can be a valuable nutrient management tool but usually requires additional technical information regarding the timing, form, and amount of N addition to synchronize soil N availability with crop N requirements (McIsaac et al., 1985).

Ambient levels of nitrate in irrigation water contribute to the inorganic N pool in soil, but fertilizer credit given for waterborne nitrate may be overestimated if irrigation occurs late in the growing season, after the crop has assimilated most of the required N. Tagged-N studies indicate this is the case for nitrate water applied to corn in July and August (Francis and Schepers, 1986; Varvel et al., 1985). Therefore, fertilizer or other N sources, which are subject to leaching by spring precipitation, are frequently required to meet early season crop N requirements. Late season irrigation N applications may not be used effectively because N uptake by many field crops precedes dry matter accumulation. Although nitrate in irrigation water can be used by crops, its contribution to yield should be assessed relative to other N sources and synchronized with crop N requirements. Information about nitrate in the drainage waters associated with irrigation can be found in MacKenzie and Viets (1974).

F. Well Protection

Many groundwater contamination problems with nitrate in rural areas simply involve direct surface runoff into older, hand-dug or newer, improperly installed drilled wells (Smith et al., 1987). In such cases high ammonium contents may also be observed. Cloudy, organic matter containing well water is a good clue that surface contamination is involved. Then bacterial determination (i.e., coliforms) may be in order. This is because infantile methemoglobinemia can be a complication of enteritis (bacterial origin), even when nitrate intake is negligible (Bryson and Boeckman, 1989).

With drilled wells, the problem is usually an improper seal, which allows surface runoff into the annulus between the borehole and casing. Groundwater seepage also tends to follow the same general direction as that of surface water. To the extent possible, agricultural field, feedlot, farmstead, and septic tank drainage should be directed away from the well. In the case of feedlots, most nitrate problems occur when they are not in use. Then the surface seal breaks down, conditions become more aerobic, and nitrate production from the accumulated ammonium is initiated. Also to be avoided around wells are the storage, mixing, and loading of N fertilizer.

V. N Research Tools

A. Tracers

Nitrogen tracers provide the only definitive means for determining both the fate and the behavior of applied N in the environment (L'Annunziata and Legg,

1984; Vose, 1980). A basic assumption is that N isotopes behave in an identical manner chemically, physically, and biologically. Natural N is also assumed to have a fairly constant N-15 isotope composition. In the case of most N-15 enrichment field studies, these assumptions cause no problems. Currently, both N-15–enriched and -depleted (i.e., essentially 100 atom % N-14) fertilizer materials are available for field research.

The sensitivity of detection for stable N isotopes is relatively low because of analytical limitations and some natural variations in isotope abundance. To distinguish isotopically enriched from natural N, the N-15 abundance in the sample must lie outside the range of natural variations (approximately $0.366 \pm 0.004\%$). Consequently, the permissible dilution of N-15 enriched samples is limited to 100/0.004 or 25,000 times its weight, and that of N-15 depleted samples is much less. Various instrumentation options are now available for isotope analyses. For less exacting work (1 to 3% precision), relatively inexpensive and simple N-15 emission spectrometers are available. For high precision work ($<0.2\%$) the isotope ratio mass spectrometers can be used. In addition, highly sensitive, automated commercial mass spectrometry analysis, capable of analyzing 200 or more samples per day, is now available.

Actual N-15 enrichments in fertilizer studies have varied from about 1 to nearly 100%, but are usually in the order of 5 to 10% for most routine fertilizer efficiency studies. For those field studies involving N cycling (e.g., mineralization–immobilization, leaching, gaseous N emission) and a complete N fertilizer balance, much higher enrichment may be used to provide for the enormous dilution effect. Depending on the length and conditions of the study, from 10 to nearly 100% enrichment may be desired.

Large quantities of N-15–depleted materials are now available as a by-product of N-15 enrichment for use in field-scale studies. Practical considerations center on analytical sensitivity and isotope dilution. Generally, the use of N-15–depleted materials is confined to single-season, plant uptake studies, or those simply following the movement of applied N from soil to water.

Variations in natural N isotope abundance have been used in some N cycling and environmental studies, but the analytical and procedural problems are considerable. This is because, as noted earlier, the natural variations are extremely small and, in addition, may change during N transformations (Broadbent et al., 1980; Smith et al., 1983a). To date, the approach has not received general acceptance.

The balance sheet approach with N-15 enrichment offers the most complete way to determine the disposition of applied N in the environment. The proper experimental method depends on the resources available, the objectives, and the N transformations of interest. Due to costs, sampling problems, and spatial variability, N-15 field studies are typically confined to small plots, often <1 m^2. Sampling and spatial variability can be critical, because of N transformations associated with water content and movement. Small plots allow complete removal of the soil, thereby avoiding problems from coring and spatial variability. However, such removal does destroy the plot for future sampling. Small plots also tend to restrict fertilizer disposition studies to fibrous-rooted plants. The plots are often enclosed with a casing material to prevent lateral movement of the fertilizer, and they may be instrumented with various probes and sampling devices. Indiscriminate use of such substances should be avoided, because they may influence the N transformation and thereby vitiate the results (Smith and Young, 1984). In addition, it should be noted that some soils are more suited to N lysimetry studies than others. For example, special care should be exercised in the case of swelling clay soils, because enclosure can enhance both surface ponding and vertical movement down the lysimeter walls.

For those cases when N isotopes are too expensive to follow the potential path of soluble N (i.e., nitrate) movement in a field or watershed, a substitute tracer can be used. Bromide is the substitute tracer that has received the most attention. It moves very similarly to nitrate, has generally low natural occurrence, and is not regarded as particularly toxic to plants (Smith and Davis, 1974). Caution should be exercised, however, if livestock are confined to pastures receiving high rates of bromide (Owens et al., 1985).

B. Models

Agricultural nitrogen models represent descriptions of how soluble, particulate, and gaseous N forms are transported under field conditions. They should not violate basic physical, chemical, or biological processes but of necessity, because of various knowledge gaps, they include empirical relations. Often, the basic component involves a hydrologic model that has a submodel for N transformations. Comprehensive N models are complex, and field data necessary for adequate testing are not always available. Nevertheless, models can serve as useful basic and applied research tools to integrate multidisciplinary aspects, determine relative management change impacts, and pinpoint where specific information and understanding are lacking. In fact, limited process N models are now moving from the research to the actual field-implementation stage.

One of the early ARS models that incorporated an N transport feature was ACTMO, introduced in 1975 by Frere et al. and later expanded by Stewart et al. (1976). Further development occurred in 1980 when CREAMS, probably the most widely used and successful general agricultural chemical, field-scale model, became available (Knisel, 1980). Since then, several other models with N transport and transformation features, in some cases extensions or modifications of those just mentioned, have come on the scene. Examples include AGNPS, DRASTIC, EPIC, GLEAMS, LEACHMN, NCSOIL, NITWAT, NLEAP, NTRM, and RZWQM. Space limitations preclude a detailed consideration of the models here, so only a brief representation follows. For additional information on N models involving soil–plant–water relationships, reference is made to Frissel and van Veen (1981) and DeCoursey (1989).

The Agricultural Non-Point-Source Pollution Model (AGNPS) is a single-event, cell-based model developed to simulate sediment and nutrient transport from agricultural watersheds in Minnesota (Young et al., 1987). Basic components include hydrology, sediment, and N transport. The N component is subdivided into soluble and particulate forms. Output can be examined for either a single cell, or the whole watershed. At present, use of the model is being extended beyond Minnesota.

DRASTIC represents an empirical standardized system for evaluating groundwater pollution potential through hydrogeologic settings (Aller et al., 1987). The acronym stands for the seven parameters the National Water Well Association considers the most important mappable factors controlling groundwater pollution: *d*epth to water, net *r*echarge, *a*quifer material, *s*oil, *t*opography (slope), *i*mpact of the vadose zone, and *c*onductivity of the aquifer. These factors are incorporated into a relative ranking system that uses a combination of weights and ratings to produce a numerical index. Field testing of the approach has been lacking, but recent application to the Big Spring, Iowa, groundwater basin was disappointing (Curry, 1987). Using nonparametric statistics, DRASTIC was applied to one set of nitrate data throughout the basin. There was no evidence of correlation with field data. Lack of correlation was attributed to failure of the weighting system or inappropriate methodology for the karst area involved.

Erosion Productivity Impact Calculator (EPIC) (Williams et al., 1984) was developed to determine the relationship between soil erosion and soil productivity, and it is capable of simulating long periods (e.g., >50 years). EPIC is a comprehensive model that comprises nine major divisions: hydrology, weather, erosion, nutrients, plant growth, soil temperature, tillage, economics, and plant environment control. The N nutrient section includes surface runoff, leaching, denitrification, particulate N transport, mineralization, immobilization, crop uptake, fixation, and rainfall. The whole model is operational, but additional testing is underway for the N portion. Recent field tests with the nitrate leaching portion have been encouraging (Williams and Kissel, 1990). Tests involving N uptake by plants (Godwin and Jones, 1989) indicate that N accuracy is highly dependent on growth and yield components, and how well initial soil conditions are known.

Groundwater Loading Effects of Agricultural Management Systems (GLEAMS) (Leonard et al., 1987) was developed for field-size areas to evaluate the effects of agricultural management systems on the movement of agricultural chemicals within and through the plant root zone. The model uses the physically based CREAMS, and incorporates a component for vertical flux of the chemicals. Results of model testing demonstrate that GLEAMS simulates bromide movement and leaching within the general range of field data variability. Therefore, it should also have utility for simulating potential nitrate movement and leaching.

Leaching Estimates And Chemistry Model Nitrogen (LEACHMN) is a process-based model of water and N movement, transformations, plant uptake, and N reactions in the unsaturated zone (Wagenet and Hutson, 1987). Simulations have been formulated to describe the leaching of N fertilizer. They include volatilization of NH_4, and consider N uptake by plants to be a function of root density and NH_4 and NO_3 concentration.

Nitrogen and Water (NITWAT) management (McIsaac et al., 1985; Watts and Hanks, 1978) has been developed especially for corn on sandy soils. N transformations and transport are treated in relation to crop growth under given weather and irrigation conditions. The model results for plant N uptake, N in the soil, and leaching of nitrate from the root zone were shown recently to compare favorably with field tests (Ahuja et al., 1989).

Nitrate Leaching and Economic Analysis Package (NLEAP) (Shaffer et al., 1990) is a computer application package developed to obtain estimates of potential nitrate leaching from agricultural areas and projected impacts of this leaching on associated aquifers. Management practice, soil, climate, and economic impacts are required for application. Calibration and testing of the model has proceeded, utilizing results from sites in Nebraska and Michigan.

Nitrogen Tillage and Residue Management (NTRM) (Shaffer and Larson, 1987) is a comprehensive model that emphasizes management of nitrogen sources at the soil surface in conventional and reduced-tillage systems. N transformations and transport are processes treated in detail with use of the NCSOIL submodel, which contains both active and passive N pools. Analysis to date has been mainly with laboratory incubation data, but field testing is now underway.

Root Zone Water Quality Management (RZWQM) (Rojas et al., 1988) is a comprehensive model that includes macropore flow and a description of the N cycle. It requires a large soil physical and chemical data base, has a semi–user-friendly input format, and attempts to use an expert systems approach. Development is in progress, and the model is expected to have utility for comparing alternative management practices and their susceptibility for groundwater contamination.

Accurately modeling the transformations and transport of N in the environment remains a formidable and challenging task, because of complexity of the behavior of N in the environment. Its valence can range from +5 to −3; it exists in both organic and inorganic forms; and it occurs in gaseous, liquid, and solid states. Moreover, it is highly influenced by temperature, water, microbial, and plant processes. Consequently, any N model will continue to contain a certain amount of empiricism and estimation, especially in the case of comprehensive N models. Perhaps those that will prove to have the most field application are the ones that rely on only one or two N processes, such as leaching or volatility. As implementation of best agricultural N management practices proceeds, the need for applying comprehensive N pollution models may diminish. This is because less N should be available for pollution. Consequently, the primary focus on N models could shift from an environmental to a fertilizer economics perspective.

VI. Research Needs

During our preparation of this review various N research areas requiring additional attention surfaced or became more obvious. Perhaps the most striking general need is that for an integrated, or holistic, field-based approach to assessing and managing N losses to the environment. Usually, field watershed studies have tended to consider only one specific avenue of N loss, such as leaching or runoff. Rarely are two or more avenues considered simultaneously, which makes it difficult to assemble a comprehensive picture of N behavior. Consequently, a general recommendation is that representative watersheds be established and instrumented in certain major land resource areas where N fertilizer usage is concentrated. Instrumentation would involve devices for studying mechanisms and processes involved with N losses to surface runoff, groundwater, any tile drainage, and the atmosphere. The watersheds could be managed to reflect both rotation and monoculture cropping practices of the area, with a goal of maintaining acceptable yields and minimizing N leakage to the environment. We bring into focus several specific research needs, some of which are being actively pursued by N researchers. No particular attempt has been made to place these needs in any type of national priority, because they vary somewhat according to area and management.

1. Assess and manage any deleterious N effects of conservation tillage on water quality and the atmosphere, under various soil, climatic, and management conditions.
2. Increase investigation of the role of soil macropores in leaching of N. When necessary, devise N-leaching submodels that take some quantitative account of macropore influence.
3. Increase N balance-field studies with tagged N fertilizers, crop residues, and/or animal wastes to assess specific avenues and extent of N leakage to the environment. Gear these studies to current cropping and management schemes.
4. Continue development of soil and tissue N tests. Integrate these tests with the mass balance, N fertilizer management approach, and crop growth models to ensure timely application of N.
5. Evaluate recent developments in equipment design regarding precise timing, distribution, injection, and placement of N fertilizer type, as an aid to minimizing N losses to the environment.
6. Direct nitrification inhibitor and N stabilizer research to provide compounds and formulations geared to specific application and management techniques. The mode of action should be known so that chemical formulation can be made with predictable specificity and efficacy.
7. Tailor N fertilizer management techniques to uniform soil areas rather than whole fields, incorporating appropriate Geographic Information Systems (GIS).
8. Continue work to quantify and predict applied and soil N behavior, but center N modeling approaches more on implementation and user application, to ensure proper management decisions.
9. Improve information regarding suitable rotations, cover, and N scavenging crops to alleviate nitrate build-up in soil profiles.
10. Obtain additional information on buffer strip use. Current methods use slope as the primary controlling factor, but effects of soil, vegetation, and receiving area require study.

11. Direct more attention to the role of soil fauna (e.g., earthworms, arthropods, protozoa) regarding N behavior.
12. Determine both nitrate and bacterial (e.g., coliform) contents in studies involving N pollution of potable waters. This is recommended because infantile methemoglobinemia can be a complication of enteritis (bacterial origin), even when nitrate intake is negligible (Bryson and Boeckman, 1989).
13. Continue efforts to modify plant N metabolic processes through genetic manipulation, to improve N efficiency.
14. Develop furrow irrigation techniques to improve uniformity of water applied and facilitate fertigation management practices.
15. Integrate microbial N cycling, fertilizer management practices, and crop N requirements to improve N use efficiency by synchronizing N availability and crop requirements.
16. Develop dynamic models that will predict nitrate movement as influenced by fertilizer N placement, irrigation application systems, and plant root uptake pattern.
17. Obtain additional information on managing nitrate losses associated with tile-drained land.

References

Ahuja, L.R., D.L. Martin, and S.J. Smith. 1989. Modeling for best management of water and nutrients in sandy soils. Abstracts Int. Symp. Managing Sandy Soils. (Jodhpur, India) Part I:398–401.

Alberts, E.E., G.E. Schuman, and R.E. Burwell. 1978. Seasonal runoff losses of nitrogen and phosphorus from Missouri Valley loess watersheds. J. Environ. Qual. 7:203–208.

Aller, L., T. Bennett, J.H. Lehr, R.J. Petty, and G. Hackett. 1987. DRASTIC: A standardized system for evaluating ground water pollution potential using hydrogeologic settings. EPA/600/2-87/035. Kerr Environ. Res. Lab., Ada, OK.

Anderson, I.C., and M.A. Poth. 1987. Emissions of nitric oxide, nitrous oxide and carbon dioxide from burned and unburned chaparral soils. Agron. Abst. p. 177.

Anonymous. 1989. Nonstop nitrogen tests. Farm J. 113(6):18.

Avnimelech, Y., and M. Laher. 1977. Ammonia volatilization from soils: equilibrium considerations. Soil Sci. Soc. Am. J. 41:1080–1084.

Baker, J.L. 1985. Management practices to reduce farm chemical losses with agricultural drainage. In Perspectives on Nonpoint Source Pollution. pp. 467–470. EPA 440/5-85-001, U.S. Government Printing Office, Washington, D.C.

Baker, J.L., T.S. Colvin, S.J. Marley, and M. Dawlebeit. 1985. Improved fertilizer management with a point-injection applicator. ASAE Paper 85-1516, St. Joseph, MI.

Below, F.E., and L.E. Gentry. 1987. Effect of mixed N nutrition on nutrient accumulation, partitioning and productivity of corn. J. Fert. Issues 4:79–85.

Below, F.E., R.J. Lambert, and R.H. Hageman. 1984a. Foliar applications of nutrients on maize. I. Yield and N content of grain and stover. Agron. J. 76:773–777.

Below, F.E., R.J. Lambert, and R.H. Hageman. 1984b. Foliar application of nutrients on maize. II. Physiological responses. Agron. J. 76:777–784.

Berg, W.A., S.J. Smith, and G.A. Coleman. 1988. Management effects on runoff, soil, and nutrient losses from highly erodible soils in the Southern Plains. J. Soil Water Cons. 43:407–410.

Blackmer, A.M., and J.M. Bremner. 1977. Gas chromatographic analysis of soil atmospheres. Soil Sci. Soc. Am. J. 41:908–912.

Blackmer, A.M., and M.E. Cerrato. 1986. Soil properties affecting formation of nitric oxide by chemical reactions of nitrite. Soil Sci. Soc. Am. J. 50:1215–1218.

Blackmer, A.M., S.G. Robbins, and J.M. Bremner. 1982. Diurnal variability in rate of emission of nitrous oxide from soils. Soil Sci. Soc. Am. J. 46:937–942.

Bock, B.R., and D.E. Kissel (eds.). 1988. Ammonia volatilization from urea fertilizers. Bulletin Y-206. TVA, Muscle Shoals, AL.

Bowers, S.A., S.J. Smith, and V.L. Hauser. 1975. Influence of coarse textured fertilizer bands, surface ridges, and vertical straw mulches on nitrate leaching in soil. ARS-S-56. U.S. Department of Agriculture, New Orleans, LA.

Bremner, J.M. 1965. Nitrogen availability indexes. In C.A. Black (ed.). Methods of Soil Analysis. Part 2. Agronomy 9:1324–1345.

Broadbent, F.E. 1978. Mineralization, immobilization and nitrification. In Proceedings National Conference on Management of Nitrogen in Irrigated Agriculture. Sacramento, Calif. May 15–18, 1978. P.F. Pratt (ed.). pp. 109–134. Dept. of Soil and Environmental Sciences, Univ. Calif., Riverside, CA.

Broadbent, F.E., R.S. Rauschkolb, K.A. Lewis, and G.Y. Chang. 1980. Spatial variability in some virgin and cultivated soils. Soil Sci. Soc. Am. J. 44:524–527.

Bryson, D.D., and O.C. Boeckman. 1989. Water well methemoglobinemia and water with less than 22.6 mg NO_3-N/L. Preprints 197th American Chemical Society National Meeting (Dallas, Tx). Vol. 29(1):18–19.

Cabrera, M.L., and D.E. Kissel. 1988. Potentially mineralizable nitrogen in disturbed and undisturbed soil samples. Soil Sci. Soc. Am. J. 52:1010–1015.

Cady, F.B., and W.V. Bartholomew. 1960. Sequential products of anaerobic denitrification of Norfolk soil material. Soil Sci. Soc. Am. Proc. 24:477–482.

Campbell, C.A., Y.W. Jame, and R. DeJong. 1988. Predicting net nitrogen mineralization over a growing season: Model verification. Can. J. Soil Sci. 68:537–552.

Carlone, M.R., and W.A. Russell. 1987. Response to plant densities and nitrogen levels for four maize cultivars from different eras of breeding. Crop Sci. 27:465–470.

Chalk, P.M., and C.J. Smith. 1983. Chemodenitrification. In J.R. Freney and J.R. Simpson (eds). Gaseous Loss of Nitrogen from Plant-Soil Systems. pp. 65–89. Martinus Nijhoff and Dr. W. Junk Pub., The Hague.

Clark, F.E. 1962. Losses of nitrogen accompanying nitrification. Intern. Soil Sci. Transactions, New Zealand. pp. 477–482.

Clark, F.E., and W.E. Beard. 1960. Influence of organic matter on volatile loss of nitrogen from soil. Inter. Congr. Soil Sci. 7th Congr. Madison, WI, Comm. III. 501–508.

Clark, F.E., W.E. Beard, and D.H. Smith. 1960. Dissimilar nitrifying capacities of soils in relation to losses of applied nitrogen. Soil Sci. Soc. Am. Proc. 24:50–54.

Cole, G.W., L. Lyles, and L.J. Hagen. 1982. A simulation model of daily wind erosion soil loss. ASAE Paper 82-2575. St. Joseph, MI.

Curry, D.S. 1987. Assessment of empirical methodologies for predicting ground water pollution from agricultural chemicals. In Fairchild, D.M. (ed.) Ground Water Quality and Agricultural Practices. pp. 227–245. Lewis Pub., Chelsea, MI.

Davidson, J.M., D.A. Graetz, P.S.C. Rao, and H.M. Selim. 1978. Simulation of nitrogen movement, transformations, and uptake in plant rootzone. EPA-600-3-78-029.

Environmental Research Laboratory, U.S. Environmental Protection Agency, Athens, GA.
DeCoursey, D.G. (ed.). 1989. Proc. International Symposium on Water Quality Modeling of Agricultural Non-Point Sources. Utah State Univ., Logan.
De Datta, S.K. 1981. Principles and practices of rice production. John Wiley & Sons, New York.
Delany, A.C., D.R. Fitzjarrald, D.H. Lenschow, R. Pearson, Jr., G.J. Wendel, and B. Woodruff. 1986. Direct measurements of nitrogen oxides and ozone fluxes over grassland. J. Atmospheric Chem. 4:429–444.
Denmead, O.T. 1979. Chamber system for measuring nitrous oxide emission from soil in the field. Soil Sci. Soc. Am. J. 43:89–95.
Denmead, O.T. 1983. Micrometeorological methods for measuring gaseous losses of nitrogen in the field. In J.R. Freney and J.R. Simpson (eds.) Gaseous Loss of Nitrogen from Plant Soil Systems. pp. 133–157. Martinus Nijhoff/Dr. W. Junk Publishers, The Hague.
Denmead, O.T., J.R. Freney, and J.R. Simpson. 1976. A closed ammonia cycle within a plant canopy. Soil Biol. Biochem. 8:161–164.
Denmead, O.T., J.R. Freney, and J.R. Simpson. 1982. Dynamics of ammonia volatilization during furrow irrigation of maize. Soil Sci. Soc. Am. J. 46:149–155.
DeSmedt, F., and P.J. Wierenga. 1978. Approximate analytical solution for solute flow during infiltration and redistribution. Soil Sci. Soc. Am. J. 42:407–412.
Dick, W.A., W.M. Edwards, and F. Haghiri. 1986. Water movement through soil to which no-tillage cropping practices have been continuously applied. Proc. Agricultural Impacts on Ground Water. pp. 243–252. Natl. Water Well Assoc., Dublin, OH.
Doran, J.W., and M.S. Smith. 1987. Organic matter management and utilization of soil and fertilizer nutrients. In Soil Fertility and Organic Matter as Critical Components of Production Systems. pp. 53–72. Soil Sci. Soc. Am. Spec. Pub. No. 19, Madison, WI.
Dowdell, R.J., J.R. Burford, and R. Cress. 1979. Losses of nitrous oxide dissolved in drainage water from agriculture land. Nature 278:342–343.
Dunigan, E.P., R.A. Phelan, and C.L. Modart, Jr. 1976. Surface runoff losses of fertilizer elements. J. Environ. Qual. 5:339–342.
Duvick, D.N., and A.J. Cavalieri. 1987. Do hybrids differ in response to differential levels of nitrogen fertilizer? Proc. of the Seventeenth North Central Extension-Industry Soil Fertility Workshop. pp. 95–100. St. Louis, MO.
Edwards, W.M., L.B. Owens, D.A. Norman, and R.K. White. 1980. A setting basin–grass filter system for managing runoff from a paved beef feedlot. In Livestock Waste: A Renewable Resource. pp. 265–267. Proc. 4th Intern. Symp. on Livestock Waste. Amer. Soc. Agr. Engr., St. Joseph, MI.
Edwards, W.M., L.B. Owens, and R.K. White. 1983. Managing runoff from a small, paved beef feedlot. J. Environ. Qual. 12:281–286.
Elliott, R.L., and W.R. Walker. 1982. Field evaluation of furrow infiltration and advance functions. Trans. ASAE. 25:396–400.
Emerson, K., R.C. Russo, R.E. Lund, and R.V. Thurtson. 1975. Aqueous ammonia equilibrium calculations: Effect of pH and temperature. J. Fish. Res. Board Can. 32: 2379–2383.
Fenn, L.B., and D.E. Kissel. 1973. Ammonia volatilization from surface applications of ammonium compounds on calcareous soils; 1. General theory. Soil Sci. Soc. Am. Proc. 37:855–859.

Fenn, L.B., R.M. Taylor, and J.E. Matocha. 1981. Ammonia losses from surface-applied nitrogen fertilizer as controlled by soluble calcium and magnesium: General theory. Soil Sci. Soc. Am. J. 45:777–781.

Follett, R.F., J.W.B. Stewart, and C.V. Cole (eds.). 1987. Soil fertility and organic matter as critical components of production systems. Soil Sci. Soc. Am. Spec. Pub. 19. Amer. Soc. Agronomy, Madison, WI.

Francis, D.D., and J.S. Schepers. 1986. Nitrogen uptake by corn from fertilizer and irrigation sources. Agron. Abstr. p. 199.

Freney, J.R., and A.S. Black. 1988a. Importance of ammonia volatilization as a loss process. pp. 156–173. In J.R. Wilson (ed.). Advances in Nitrogen Cycling in Agricultural Ecosystems. C.A.B. International. Wallingford, England.

Freney, J.R., and E. Humphreys. 1988b. Gaseous losses of nitrogen from Agricultural Systems in Temperate Australia. In Nitrogen Cycling in Temperate Agriculture Systems. Vol. II. pp. 297–315. P.E. Bacon et al. (eds). Australian Soc. Soil Sci.

Freney, J.R., A.R. Leuning, J.R. Simpson, O.T. Denmead, and W.A. Muirhead. 1985. Estimating ammonia volatilization from flooded rice fields and by simplified techniques. Soil Sci. Soc. Am. J, 49:1049–1054.

Freney, J.R., and J.R. Simpson (eds.). 1983. Gaseous Loss of Nitrogen from Plant Soil Systems. Developments in Plant and Soil Sciences, Vol. 9. pp. 1–313. Martinus Nijhoff/Dr. W. Junk Publishers, The Hague.

Freney, J.R., J.R. Simpson, and O.T. Denmead. 1983. Volatilization of ammonia. In J.R. Freney, and J.R. Simpson (eds.). Gaseous Loss of Nitrogen from Plant-Soil Systems. pp. 1–32. Martinus Nijhoff/Dr. W. Junk Publishers, The Hague.

Frere, M.H., C.A. Onstad, and H.N. Holtan. 1975. ACTMO, An Agricultural Chemical Transport Model. ARS-H-3. U.S. Department of Agriculture, Hyattsville, MD.

Frissel, M.J., and J.A. van Veen (eds.) 1981. Simulation of nitrogen behavior of soil-plant systems. Centre for Agric. Pub. and Documentation. Wageningen, Netherlands.

Frye, W.W., R.L. Blevins, M.S. Smith, and S.J. Corak. 1988. Role of annual legume cover crops in efficient use of water nitrogen. In W.L. Hargrove (ed.). Cropping Strategies for Efficient Use of Water and Nitrogen. pp. 129–154. Am. Soc. Agron. Spec. Pub. No. 51.

Galbally, I.E., J.R. Freney, W.A. Muirhead, J.R. Simpson, A.C.F. Trevitt, and P.M. Chalk. 1987. Emission of nitrogen oxides (NO_x from a flooded soil fertilized with urea): Relation to other nitrogen loss processes. J. Atm. Chem. 5:343–365.

Galbally, I.E., and C.R. Roy. 1978. Loss of fixed nitrogen from soils by nitric oxide exhalation. Nature 275:734–735.

Gambrell, R.P., J.W. Gilliam, and S.B. Weed. 1975. Nitrogen losses from soils of the North Carolina Coastal Plain. J. Environ. Qual. 4:317–323.

Gianello, C., and J.M. Bremner. 1988. A rapid steam distillation method of assessing potentially available organic nitrogen in soil. Commun. Soil Sci. Plant Anal. 19:1551–1568.

Gilley, J.R., and L.N. Mielke. 1980. Conserving energy with low-pressure center pivots. J. Irrig. Drainage Div. ASCE 106:49–59.

Gilliam, J.W., and G.D. Hoyt. 1987. Effect of conservation tillage on fate and transport of nitrogen. In Logan, T.J. et al. (eds.) Effects of Conservation Tillage on Groundwater Quality. Ch. 13, pp. 217–240. Lewis Pub. Co., Chelsea, MI.

Gilliam, J.W., T.J. Logan, and F.E. Broadbent. 1985. Fertilizer use in relation to the environment. In O.P. Englestad (ed.) Fertilizer Technology and Use, 3rd ed. pp. 561–588. Soil Sci. Soc. Amer., Madison, WI.

Godwin, D.L., and C.A. Jones. 1990. Nitrogen dynamics in soil-plant systems–The CERES-N Models. In R.J. Hanks and J.T. Ritchie (eds.) Amer. Soc. Agron. Monograph, Madison, WI. (In Press)

Hageman, R.H. 1978. Effect of form of nitrogen on plant growth. In J.J. Meisinger et al. (eds.). Nitrification Inhibitors–Potentials and Limitations. pp. 46–62. Am. Soc. Agron. Spec. Pub. No. 38.

Hagen, L.J., and L. Lyles. 1985. Amount and nutrient content of particles produced by soil aggregate abrasions. In Erosion and Soil Productivity. pp. 117–129. ASAE Pub. 8-85. St. Joseph, MI.

Hallberg, G.R., J.L. Baker, and G.W. Randall. 1986. Utility of tile-line effluent studies to evaluate the impact of agricultural practices on ground water. In Proc. Agricultural Impacts on Ground Water–A Conference. pp. 298–328. Water Well Journ. Pub. Co., Dublin, OH.

Hargett, N.L., J.T. Berry, and S.L. McKinney. 1988. Commercial Fertilizers. National Fertilizer Development Center, Tennessee Valley Authority. Muscle Shoals, AL.

Hauck, R.D. (ed.). 1984. Nitrogen in Crop Production. American Soc. Agronomy, Madison, WI.

Hauck, R.D., and D.R. Bouldin. 1961. Distribution of isotopic nitrogen gas during denitrification. Nature 191:871–872.

Hauck, R.D., S.W. Melsted, and P.E. Yankwich. 1958. Use of N-isotope distribution in nitrogen gas in the study of denitrification. Soil Sci. 86:287–291.

Heenan, D.P., and P.E. Bacon. 1987. Effect of nitrogen fertilizer timing on crop growth and nitrogen use efficiency by different rice varieties in southeastern Australia. In Efficiency of Nitrogen Fertilizers for Rice. Proc. of the meeting of the International Network on Soil Fertility and Fertilizer Evaluation for Rice, Griffith, New South Wales, Australia, 10–16 April 1985. pp. 97–105. International Rice Institute, Los Banos, Laguna, Philippines.

Hergert, G.W., and R.A. Wiese. 1978. Performance of nitrification inhibitors in the midwest (west). In J.J. Meisinger et al. (eds.). Nitrification Inhibitors–Potentials and Limitations. pp. 89–105. Am. Soc. Agron. Spec. Pub. No. 38.

Herlihy, M. 1979. Nitrogen mineralization in soils of varying texture, moisture and organic matter. I. Potential and experimental values in fallow soils. Plant Soil 53: 255–267.

Hesterman, O.B. 1988. Exploiting forage legumes for nitrogen contribution in cropping systems. In W.L. Hargrove (ed.). Cropping Strategies for Efficient Use of Water and Nitrogen. pp. 155–166. Am. Soc. Agron. Spec. Pub. No. 51.

Howard-Williams, C., S. Pickmere, and J. Davies. 1986. Nutrient retention and processing in New Zealand streams: The influence of riparian vegetation. New Zealand Agric. Sci. 20:110–114.

Hutchinson, G.L., and R.F. Follett. 1986. Nitric oxide emissions from fallow soil as influenced by tillage treatments. Agron. Abstr., p. 180.

Hutchinson, G.L., and A.R. Mosier. 1981. Improved soil cover method for field measurements of nitrous oxide fluxes. Soil Sci. Soc. Am. J. 45:311–316.

Hutchinson, G.L., A.C. Tortoso, and W.D. Guenzi. 1987. Nitric and nitrous oxide emissions from NH_4^+–amended soil. Agron. Abstr., p. 185.

Kanwar, R.S., J.L. Baker, and J.M. Laflen. 1985. Nitrate movement through the soil profile in relation to tillage systems and fertilizer application methods. Trans. ASAE 28:1802–1807.

Keeney, D.R. 1982a. Nitrogen management for maximum efficiency and minimum pollution. Agronomy 22:605–649.

Keeney, D.R. 1982b. Nitrogen—Availability indexes. Agronomy 9:711–733.

Keeney, D.R. 1986a. Sources of nitrate to groundwater. CRC Critical Reviews in Environmental Control. 16(3):257-304.

Keeney, D.R. 1986b. Inhibition of nitrification in soils. In J.I. Prosser (ed.) Nitrification. pp. 99–115. IRL Press, Oxford.

Kemper, W.D., B.J. Ruffing, and J.A. Bondurant. 1982. Furrow intake rates and water management. Trans. ASAE 25:333–343.

Kissel, D.E., C.W. Richardson, and E. Burnett. 1976. Losses of nitrogen in surface runoff in the Blackland Prairie of Texas. J. Environ. Qual. 5:288–293.

Knisel, W.G. (ed.). 1980. CREAMS: A field scale model for chemicals, runoff, and erosion from agricultural management systems. USDA Conservation Res. Rep. 26. U.S. Government Printing Office, Washington, D.C.

L'Annunziata, M.F., and J.O. Legg (ed.). 1984. Isotopes and Radiation in Agricultural Sciences. Academic Press, London.

Legg, J.O., and J.J. Meisinger. 1982. Soil nitrogen budgets. In F.J. Stevenson (ed.). Nitrogen in Agricultural Soils. Agron. 22:503–566. Amer. Soc. Agron., Madison, WI.

Lehman, O.R., and L.R. Ahuja. 1985. Interflow of water and tracer chemical on sloping field plots with exposed seepage faces. J. Hydrology 76:307–317.

Lemon, E. 1978. Critique of "soil and other sources of nitrous oxide," nitrous oxide (N_2O) exchange of the land surface. In D.R. Nielsen and J.G. MacDonald (eds.). Nitrogen in the Environment. Vol. 1. pp. 493–521. Academic Press, New York.

Leonard, R.A., W.G. Knisel, and D.A. Still. 1987. GLEAMS: Groundwater loading effects of agricultural management systems. Trans. ASAE 30:1403–1418.

Letey, J., J.W. Blair, D. Devitt, L.J. Lund, and P. Nash. 1977. Nitrate—nitrogen in effluent from agricultural tile drains in California. Hilgardia 45:289–319.

Linn, D.M., and J.W. Doran. 1984. Effect of water-filled pore space on carbon dioxide and nitrous oxide productions in tilled and non-tilled soils. Soil Sci. Soc. Am. J. 48: 1267–1272.

Logan, T.J., G.W. Randell, and D.R. Timmons. 1980. Nutrient content of tile drainage from cropland in the north central region. Research Bulletin 1119. Ohio Agric. Res. Develop. Center, Wooster, OH.

Lowrance, R.R., R.L. Todd, and L.E. Asmussen. 1983. Waterborne nutrient budgets for the riparian zone of an agricultural watershed. Agric. Ecosystems and Environ. 10: 371–384.

Lowrance, R.R., R.L. Todd, and L.E. Asmussen. 1984. Nutrient cycling in an agricultural watershed. 1. Phreatic movement. J. Environ. Qual. 13:22–27.

MacKenzie, A.J., and F.G. Viets, Jr. 1974. Nutrients and other chemicals in agricultural drainage waters. In J. van Schilfgaarde. (ed.). Drainage for Agriculture. Agronomy 17:489–511.

Magdoff, F.R., D. Ross, and J. Amadon. 1984. A soil test for nitrogen availability to corn. Soil Sci. Soc. Am. J. 48:1301–1304.

Mathers, A.C., B.A. Stewart, and B. Blair. 1975. Nitrate-nitrogen removal from soil profiles by alfalfa. J. Environ. Qual. 4:403–405.

McClenahan, E.J., and R. Killorn. 1988. Relationship between basal corn stem nitrate N content at V6 growth stage and grain yield. J. Prod. Agric. 1:322–326.

McIsaac, G.F., D.L. Martin, and J.S. Schepers. 1984. Nitrate leaching in sandy soil. ASAE Paper No. 2610. St. Joseph, MI.

McIsaac, G.F., D.L. Martin, and D.G. Watts. 1985. Users guide to NITWAT, a nitrogen and water management tool. Dept. Agr. Eng., Univ. of Neb., Lincoln.

Meisinger, J.J. 1984. Evaluating plant-available nitrogen in soil–crop systems. In R.D.

Hauck (ed.). Nitrogen in Crop Production. pp. 391–441. Amer. Soc. Agron., Madison, WI.
Menzel, R.G. 1980. Enrichment ratios for water quality modeling. In W.G. Knisel (ed.). CREAMS: A Field Scale Model for Chemicals, Runoff, and Erosion from Agricultural Management Systems. pp. 486–492. USDA-ARS Conservation Research Report 26. U.S. Government Printing Office, Washington, D.C.
Menzel, R.G., S.J. Smith, and N.H. Welch. 1986. Effect of impoundments on nutrient concentrations. Proc. 4th Federal Interagency Sedimentation Conference. Vol. 2 (7):21–30.
Miller, D.E., J.S. Aarstad, and R.G. Evans. 1987. Control of furrow erosion with crop residues and surge flow irrigation. Soil Sci. Soc. Am. J. 51:421–425.
Mitchell, W.H. 1981. Subsurface irrigation and fertilization of field corn. Agron. J. 73:913–916.
Molina, J.A.E., C.E. Clapp, and W.E. Larson. 1980. Potentially mineralizable nitrogen in soil: The simple exponential model does not apply for the first 12 weeks of incubation. Soil Sci. Soc. Am. J. 44:442–443.
Mosier, A.R., and O. Heinemeyer. 1985. Current methods used to estimate N_2O and N_2 emissions from field soils. In Han L. Golterman (ed.). Denitrification in the Nitrogen Cycle. pp. 79–99. Plenum Pub. Corp., New York.
Mosier, A.R., and L. Mack. 1980. Gas chromatographic system for precise, rapid analysis of nitrous oxide. Soil Sci. Soc. Am. J. 44:1121–1123.
Mulvaney, R.L. 1984. Determination of ^{15}N-labeled dinitrogen and nitrous oxide with triple-collector mass spectrometers. Soil Sci. Soc. Am. J. 48:690–692.
Mulvaney, R.L., and C.W. Boast. 1986. Equations for determination of nitrogen-15 labeled dinitrogen and nitrous oxide by mass spectrometry. Soil Sci. Soc. Am. J. 50:360–363.
Mulvaney, R.L., and L.T. Kurtz. 1984. Evolution of dinitrogen and nitrous oxide from nitrogen-15 fertilized soil cores subjected to wetting and drying cycles. Soil Sci. Soc. Am. J. 48:596–602.
Naney, J.W., W.A. Berg, S.J. Smith, and G.A. Sample. 1986. Assessment of ground water quality in saline seeps. In Proc. Agricultural Impacts on Ground Water—A Conference. pp. 274–285. Water Well Journal Pub. Co., Dublin, OH.
Neibling, W.H., and E.E. Alberts. 1979. Composition and yield of soil particles transported through soil strips. ASAE Paper No. 79-2065. St. Joseph, MI.
Nelson, D.W., and J.M. Bremner. 1969. Factors affecting chemical transformations of nitrite in soils. Soil Biol. Biochem. 1:229–239.
Nelson, D.W., and J.M. Bremner. 1970. Gaseous products of nitrite decomposition in soils. Soil Biol. Biochem. 2:203–215.
Nelson, D.W., and D.M. Huber. 1978. Performance of nitrification inhibitors in the midwest (east). In J.J. Meisinger et al. (eds.). Nitrification inhibitors—Potentials and Limitations. pp. 75–88. Am. Soc. Agron. Spec. Pub. No. 38.
Nielsen, D.R., J.W. Biggar, and P.J. Wierenga. 1982. Nitrogen transport processes in soil. In F.J. Stevenson (ed.). Nitrogen in Agricultural Soils. Agronomy 22:423–448.
Nielsen, D.R., and J.G. MacDonald (eds.). 1978. Nitrogen in the Environment. Vol. 1, Academic Press, New York.
Nommik, H. 1956. Investigations on denitrification in soil. Acta Agric. Scand. 6:195–228.
Ohlander, C.A. 1976. Defining the sediment trapping characteristics of a vegetative buffer. Special case: Road erosion. In Proc. Third Federal Interagency Sedimentation Conf. pp. 2-77–2-81. U.S. Government Survey, Denver, CO.
Olson, R.A. 1986. Agricultural practices for minimizing nitrate content of ground water.

In Proc. Agricultural Impacts on Ground Water—A Conference. pp. 428–444. Water Well Journal Publish Co., Dublin, OH.
Olson, R.A., E.C. Seim, and J. Muir. 1973. Influence of agricultural practices on water quality in Nebraska: A survey of streams, groundwaters and precipitation. Water Resources Bull. 9:301–311.
Olsen, R.J., R.F. Hensler, O.J. Attoe, S.A. Witzel, and L.A. Peterson. 1970. Fertilizer nitrogen and crop rotation in relation to movement of nitrate nitrogen through soil profiles. Soil Sci. Soc. Am. Proc. 34:448–452.
Olsen, S.R. 1986. Role of ammonium nutrition in higher yields. Proc. Great Plains Soil Fertility Workshop, March 4–5. Denver, CO, pp. 6–14.
Omernik, J.M., A.R. Abernathy, and L.M. Male. 1981. Stream nutrient levels and proximity of agricultural and forest lands to streams: Some relationships. J. Soil Water Conser. 36:227–231.
Onken, A.B. 1978. Performance of nitrification inhibitors in the Southwest. In J.J. Meisinger et al. (eds.). Nitrification Inhibitors—Potentials and Limitations. pp. 119- 129. Am. Soc. Agron. Spec. Pub. No. 38.
Organization for Economic Cooperation and Development. 1982. Eutrophication of Waters: Monitoring, Assessment, and Control. O.E.C.D. Paris, France.
Owens, L.B. 1987. Nitrate leaching losses from monolith lysimeters as influenced by nitrapyrin. J. Environ. Qual. 16:34–38.
Owens, L.B., R.W. Van Keuren, and W.M. Edwards. 1985. Groundwater quality changes resulting from a land surface bromide application to a pasture. J. Environ. Qual. 14: 543–548.
Oyanedel, C., and J. Rodriguez. 1977. Estimacion de la mineralizacion del n en el suelo. Ciencia Investigacion Agraria 4:33–44.
Papendick, R.I., L.F. Elliott, and J.F. Power. 1987. Alternative production systems to reduce nitrate in ground water. Amer. J. Alternative Agric. 2:19–28.
Papendick, R.I., and J.C. Engibous. 1978. Performance of nitrification inhibitors in the Northwest. In J.J. Meisinger et al. (eds.). Nitrification Inhibitors—Potentials and Limitations. pp. 107–117. Am. Soc. Agron. Spec. Pub. No. 38.
Peterson, G. 1989. Geographic Information Systems. Proc. Research Issues in Soil Erosion Workshop, Univ. of Minn., St. Paul.
Pionke, H.B., and R.R. Lowrance. 1990. Fate of nitrate in subsurface drainage water. In Follett, R.F. (ed.). Managing nitrogen for ground water quality and farm profitability. Soil Sci. Soc. Amer., Madison, WI.
Porter, L.K. 1969. Gaseous products produced by anaerobic reaction of sodium nitrite with oxime compounds and oximes synthesized from organic matter. Soil Sci. Soc. Am. Proc. 33:696–702.
Power, J.F., and J.W. Doran. 1984. Nitrogen use in organic farming. In R.D. Hauck (ed.). Nitrogen in Crop Production. pp. 585–598. Amer. Soc. Agron., Madison, WI.
Rasmussen, R.A., J. Krasnec, and P.P. Pierotti. 1976. N_2O-analysis in the atmosphere via electron capture chromatography. Geophys. Res. Letter. 3:615–618.
Rauschkolb, R.S., T.L. Jackson, and A.I. Dow. 1984. Management of nitrogen in the Pacific States. In R.D. Hauck (ed.). Nitrogen in Crop Production. pp. 765–777. Amer. Soc. Agron., Madison, WI.
Reichenberger, L. 1989. Tools with eyes. Farm J. 113(6):16.
Reichenberger, L., and J. Russnogle. 1989. Farm by the foot. Farm J. 113(6):11–15.
Rhoads, F.M., and R.L. Stanley, Jr. 1981. Fertilizer scheduling, yield, and nutrient uptake of irrigated corn. Agron. J. 73:971–974.
Robinson, C.W., and C.E. Brockway. 1980. Sediment removal with vegetal filters. Irriga-

tion and Drainage–Today's Challenge. pp. 422–428. Proc. Irrigation and Drainage Division–Specialty Conf. Am. Soc. Civil. Engr.

Rojas, K.W., C.S. Hebson, and D.G. DeCoursey. 1988. Modeling agriculture management subject to subsurface water quality restraints. In Modeling Agricultural Forest and Rangeland Hydrology. pp. 108–116. ASAE Pub. 07-88. St. Joseph, MI.

Rolston, D.E., J. Krasnee, and D. Pierotti. 1976. Denitrification measured directly from nitrous oxide gas fluxes. Soil Sci. Soc. Am. J. 40:259–266.

Rose, C.W., F.W. Chichester, J.R. Williams, and J.T. Ritchie. 1982. Application of an approximate analytic method of computing solute profiles with dispersion in soils. J. Environ. Qual. 11:151–155.

Ryden, J.C., L.J. Lund, and D.D. Focht. 1978. Direct in-field measurement of nitrous oxide flux from soils. Soil Sci. Soc. Am. J. 42:731–737.

Ryden, J.C., and D.E. Rolston. 1983. The measurement of denitrification. In J.R. Freney and J.R. Simpson (eds.). Gaseous Loss of Nitrogen from Plant-Soil Systems. pp. 91–132. Martinus Nijhoff/Dr. W. Junk Pub., The Hague.

Sahrawat, K.L., and D.R. Keeney. 1986. Nitrous oxide emission from soils. In B.A. Stewart (ed.). Advances in Soil Science 4:103–148, Springer-Verlag, New York.

Schepers, J.S. 1988. Role of cropping systems in environmental quality: Groundwater nitrogen. In W.L. Hargrove (ed.). Cropping Strategies for Efficient Use of Water and Nitrogen. pp. 167–178. Am. Soc. Agron. Spec. Pub. No. 51.

Schepers, J.S., and F.E. Below. 1988. Influence of corn hybrids on nitrogen uptake and utilization efficiency. pp. 172–186. Proc. Forty-Second Annual Corn and Sorghum Industry Research Conference, Dec. 10–11. Chicago, IL.

Schepers, J.S., and R.D. Fox. 1989. Estimation of N budgets for crops. Chap. VIII. In R. Follett (ed.). Nitrogen Management for Groundwater Protection. pp. 221–246. Elsevier Science Pub., Amsterdam.

Schepers, J.S., and D.D. Francis. 1982. Chemical water quality of runoff from grazing land in Nebraska: I. Influence of grazing livestock. J. Environ. Qual. 11:351–354.

Schepers, J.S., D.D. Francis, and L.N. Mielke. 1985. Water quality from environmental control structures in Nebraska. J. Environ. Qual. 14:186–190.

Schepers, J.S., R.D. Lohry, R.B. Ferguson, and G.W. Hergert. 1988. Strategies to minimize nitrate in groundwater. pp. 1–9. Proc. Great Plains Soil Fertility Workshop, March 8-9. Denver, CO.

Schepers, J.S., and R. Saint-Fort. 1988. Comparison of potentially mineralized nitrogen using electroultrafiltration and four other procedures. Proc. 3rd Intl. EUF Symposium (Mannheim, Germany). 2:441–450.

Schertz, D.L., and D.A. Miller. 1972. Nitrate-N accumulation in the soil profile under alfalfa. Agronomy J. 64:660–664.

Schuman, G.E., R.E. Burwell, R.F. Piest, and R.G. Spomer. 1973. Nitrogen losses in surface runoff from agricultural watersheds in Missouri Valley loss. J. Environ. Qual. 2:297–302.

Shaffer, M.J., A.D. Halvorson, and F.J. Pierce. 1990. Nitrate leaching and economic analysis package (NLEAP): Model description and application. In Follett, R.F. (ed). Managing Nitrogen for Ground Water Quality and Farm Profitability. Soil Sci. Soc. Amer. Pub., Madison, WI.

Shaffer, M.J., and W.E. Larson (ed). 1987. NTRM, a soil-crop simulation model for nitrogen, tillage, and crop residue management. USDA Conservation Research Report. 34-1.

Sharpley, A.N., S.J. Smith, W.A. Berg, and J.R. Williams. 1985. Nutrient runoff losses as predicted by annual and monthly soil sampling. J. Environ. Qual. 14:354–359.
Sharpley, A.N., S.J. Smith, and J.W. Naney. 1987. Environmental impact of agricultural nitrogen and phosphorus use. J. Agric. Food Chem. 35:812–817.
Siegel, R.S., R.D. Hauck, and L.T. Kurtz. 1982. Determination of N_2 and application to measurement of N_2 evolution during denitrification. Soil Sci. Soc. Am. J. 46:68–74.
Slemr, F., and W. Seiler. 1984. Field measurement of NO and NO_2 emissions from fertilized and unfertilized soils. J. Atmospheric Chem. 2:1–24.
Smika, D.E., and P.H. Grabouski. 1976. Anhydrous ammonia applications during fallow for winter wheat production. Agron. J. 68:919–922.
Smith, C.J., and P.M. Chalk. 1980. Gaseous nitrogen evolution during nitrification of ammonia fertilizer and nitrite transformations in soils. Soil Sci. Soc. Am. J. 44:277–282.
Smith, D.H., and F.E. Clark. 1960. Volatile losses of nitrogen from acid or neutral soils or solutions containing nitrite and ammonium ions. Soil Sci. 90:86–92.
Smith, S.J. 1972. Relative rate of chloride movement in leaching of surface soils. Soil Sci. 114:259–263.
Smith, S.J. 1987. Soluble organic nitrogen losses associated with recovery of mineralized nitrogen. Soil Sci. Soc. Am. J. 51:1191–1194.
Smith, S.J., L.R. Ahuja, and J.D. Ross. 1984. Leaching of a soluble chemical under field crop conditions. Soil Sci. Soc. Am. J. 48:252–258.
Smith, S.J., and D.K. Cassel. 1990. Estimating nitrate leaching in soil materials. In Follett, R.F. (ed.). Managing Nitrogen for Ground Water Quality and Farm Profitability. Soil Science Soc. Amer. Pub., Madison, WI.
Smith, S.J., and R.J. Davis. 1974. Relative movement of bromide and nitrate through soils. J. Environ. Qual. 3:152–155.
Smith, S.J., J.O. Legg, and L.K. Porter. 1983a. N-15 abundances in virgin soils and their fertilized analogues. In R.R. Lowrance et al. (eds.). Proc. Intl. Symp. Nutrient Cycling in Agric. Ecosystems. pp. 315–324. Univ. of Georgia College Agric. Spec. Pub. 23.
Smith, S.J., R.G. Menzel, E.D. Rhoades, J.R. Williams, and H.V. Eck. 1983b. Nutrient and sediment discharge from Southern Plains grasslands. J. Range Mgt. 36:435–439.
Smith, S.J., J.W. Naney, and W.A. Berg. 1987. Nitrogen and ground water protection. In Fairchild, D.M. (ed.). Ground Water Quality and Agricultural Practices. pp. 367–374. Lewis Pub., Chelsea, MI.
Smith, S.J., A.N. Sharpley, W.A. Berg, G.A. Coleman, and N.H. Welch. 1988. Nutrient losses from agricultural land runoff in Oklahoma. Proc. 22th Okla. Agric. Chem. Conf. 13:23–26. Oklahoma State Univ. Pub., Stillwater, OK.
Smith, S.J., A.N. Sharpley, J.R. Williams, A.D. Nicks, and O.R. Jones. 1986. Sediment-nutrient transport in agricultural runoff. In Proc. 4th Federal Interagency Sedimentation Conference 2(7):11–20.
Smith, S.J., and L.B. Young. 1975. Distribution of nitrogen forms in virgin and cultivated soils. Soil Sci. 120:354–360.
Smith, S.J., and L.B. Young. 1984. Mineralization of soil nitrogen in the presence of different metallic and organic solids. Commun. Soil Sci. Plant Anal. 15:125–134.
Smith, S.J., L.B. Young, and G.E. Miller. 1977. Evaluation of soil nitrogen mineralization potentials under modified field conditions. Soil Sci. Soc. Am. J. 41:74–76.
Soil Conservation Service. 1984. National Handbook of Conservation Practices, U.S. Government Printing Office, Washington, D.C.

Stanford, G., C.B. England, and A.W. Taylor. 1970. Fertilizer use and water quality. ARS 41-168. USDA Publication.

Stanford, G., J.C. Legg, and S.J. Smith. 1973. Soil nitrogen availability evaluations based on nitrogen mineralization potentials of soils and uptake of labeled and unlabeled nitrogen by plants. Plant Soil. 39:113–124.

Stanford, G., and S.J. Smith. 1972. Nitrogen mineralization potentials of soils. Soil Sci. Soc. Am. Proc. 36:465–472.

Stevenson, F.J., R.M. Harrison, R. Wetselaar, and R.A. Leeper. 1970. Nitrosation of soil organic matter: III. Nature of gases produced by reaction of nitrite with lignins, humic substances, and phenolic constituents under neutral and slightly acidic conditions. Soil Sci. Soc. Am. Proc. 34:430–435.

Stevenson, F.J., and R.J. Swaby. 1964. Nitrosation of soil organic matter: I. Nature of gases evolved during nitrous acid treatment of lignins and humic substances. Soil Sci. Soc. Proc. 28:773–778.

Stewart, B.A., F.G. Viets, Jr., and G.L. Hutchinson. 1968. Agriculture's effect on nitrate pollution of groundwater. J. Soil Water Conser. 23:13–15.

Stewart, B.A., D.A. Woolhiser, W.H. Wischmeier, J.H. Caro, and M.H. Frere. 1976. Control of Water Pollution from Cropland. Vol. II: An Overview. ARS-H-5-2. U.S. Government Printing Office, Washington, D.C.

Strathouse, S.M., G. Sposito, P.J. Sullivan, and L.J. Lund. 1980. Geologic nitrogen: A potential geochemical hazard in the San Joaquin Valley, California. J. Environ. Qual. 9:54–60.

Stumpe, J.M. 1981. Ammonia volatilization from urea and urea phosphate in calcareous soils. Ph.D. thesis. Colorado State University.

Terman, G.L. 1979. Volatilization losses of nitrogen as ammonia from surface-applied fertilizers, organic amendments, and crop residues. Adv. in Agron. 31:189–223.

Tisdale, S.L., W.L. Nelson, and J.D. Beaton. 1985. Soil Fertility and Fertilizers. 4th ed. Macmillan Publishing Co., New York.

Tortoso, A.C. 1988. Contributions of eutrophic and hetertrophic nitrifiers to soil NO and N_2O emissions. M.S. thesis. Colorado State University.

Tortoso, A.C., G.L. Hutchinson, and W.D. Guenzi. 1986. Nitric and nitrous oxide emissions during nitrification and denitrification in soil. Agron. Abstr., pp. 190–191.

Touchton, J.T., and F.C. Boswell. 1978. Performance of nitrification inhibitors in the southeast. In J.J. Meisinger et al. (eds.). Nitrification Inhibitors—Potentials and Limitations. pp. 63–74. Am. Soc. Agron. Spec. Pub. No. 38.

Tucker, B.B., and L.W. Murdock. 1984. Nitrogen use in the south-central states. In Hauck, R.D. (ed.). Nitrogen in Crop Production. pp. 735–744. Amer. Soc. Agron., Madison, WI.

Tyler, D.D., and G.W. Thomas. 1977. Lysimeter measurements of nitrate and chloride losses from soil under conventional and no-tillage corn. J. Environ. Qual. 6:63–66.

Varvel, G.E., G.W. Hergert, K.D. Frank, and J.S. Schepers. 1985. Nitrogen recovery by corn from N-tagged subsoil and irrigation applied N applications. Agron. Abstr., p. 186.

Viets, F.G., Jr., and R.G. Hageman. 1971. Factors affecting the accumulations of nitrate in soil, water and plants. USDA Handbook 413. U.S. Government Printing Office, Washington, D.C.

Viets, F.R. 1975. The environmental impact of fertilizer. CRC Critical Reviews in Environmental Control. Sept., pp. 423–453.

Vlek, P.L.G., I.P.R. Fillary, and J.R. Burford. 1981. Accession, transformation, and loss of nitrogen in soils of arid regions. Plant and Soil. 58:133–174.

Vose, P.B. 1980. Introduction to Nuclear Techniques in Agronomy and Plant Biology. Pergamon Press, Oxford.

Voss, R.D., and W.D. Shrader. 1984. Rotation effects and legume source of nitrogen for corn. In D.F. Bezdicek, J.F. Power, D.R. Keeney, and M.J. Wright (eds.). Organic Farming: Current Technology and Its Role in a Sustainable Agriculture. pp. 61–68. ASA Spec. Pub. 46. Amer. Soc. Agron., Madison, WI.

Wadleigh, C.H. 1968. Wastes in relation to agriculture and forestry. U.S.D.A. Miscellaneous Pub. 1065.

Wagenet, R.J., and J.L. Hutson. 1987. LEACHM: Leaching Estimation and Chemistry Model. Continuum, Vol. 2. Water Resources Inst., Cornell Univ.

Wallach, R., W.A. Jury, and W.F. Spencer. 1988. Modeling for losses of soil applied chemicals in runoff: Lateral irrigation versus precipitation. Soil Sci. Soc. Am. J. 52:605–612.

Warrick, A.W., D.E. Meyers, and D.R. Nielsen. 1986. Geostatistical methods applied to soil science. In A. Klute (ed). Methods of Soil Analysis. Part 1. Physical and Mineralogical Methods, 2nd ed. Agronomy 9:53–82.

Watts, D.G., and R.J. Hanks. 1978. A soil-water-nitrogen model for irrigated corn on sandy soils. Soil Sci. Soc. Am. J. 42:492–499.

Whitney, D.A. 1970. Test for available nitrogen. L. 275. Coop. Extn. Ser., Kansas St. Univ., Manhattan.

Wiese, R.A., R.B. Ferguson, and G.W. Hergert. 1987. Fertilizer nitrogen best management practices. Neb. Guide No. G87-829. Univ. of Nebraska-Lincoln.

Wijler, J., and C.C. Delwiche. 1954. Investigations on the denitrifying process in soil. Plant and Soil 5:155–169.

Williams, J.R., C.A. Jones, and P.T. Dyke. 1984. A modeling approach to determine the relationship between erosion and soil productivity. Trans ASAE 27:129–144.

Williams, J.R., and D.E. Kissel. 1990. Water percolation: An indicator of N leaching potential. In Follet, R.F. (ed.). Managing Nitrogen for Groundwater Quality and Farm Profitability. Soil Sci. Soc. Amer., Madison, WI.

Woodruff, N.P., and F.H. Siddoway. 1965. A wind erosion equation. Soil Sci. Soc. Am. Proc. 29:602–608.

Young, R.A., T. Huntrods, and W. Anderson. 1980. Effectiveness of vegetated buffer strips in controlling pollution from feedlot runoff. J. Environ. Qual. 9:483–487.

Young, R.A., C.A. Onstad, D.D. Bosch, and W.P. Anderson. 1987. AGNPS, Agricultural Non-Point-Source Pollution Model. A Watershed Analysis Tool. USDA Cons. Res. Rep. 35, Natl. Tech. Info. Ser., Springfield, VA.

Zobeck, T.M., and D.W. Fryrear. 1986. Chemical and physical characteristics of wind-blown sediment. II. Chemical characteristics and total soil and nutrient discharge. Trans. ASAE 29:1037–1041.

Assessing and Modifying Ca and Al Levels in Acid Subsoils

J.M. McCray and M.E. Sumner

I. Introduction

Subsoil acidity is a major problem in many parts of the world, particularly on highly weathered soils. In addition, subsoil acidity has been reported in some areas where the acidity is derived from the parent material rather than advanced weathering processes. The restricted rooting zone that results from excessive aluminum (Al) or insufficient calcium (Ca) in these soils becomes particularly acute during dry periods because the moisture in the subsoil is beyond the reach of roots. These soils are usually located in areas receiving adequate annual rainfall, but with a growing season during which drought conditions frequently occur. Much of the rainfall received during the growing season in these areas comes in the form of intense thunderstorms, resulting in a higher percentage of water being lost as runoff, which further exacerbates the problem.

Advances in Soil Science, Volume 14

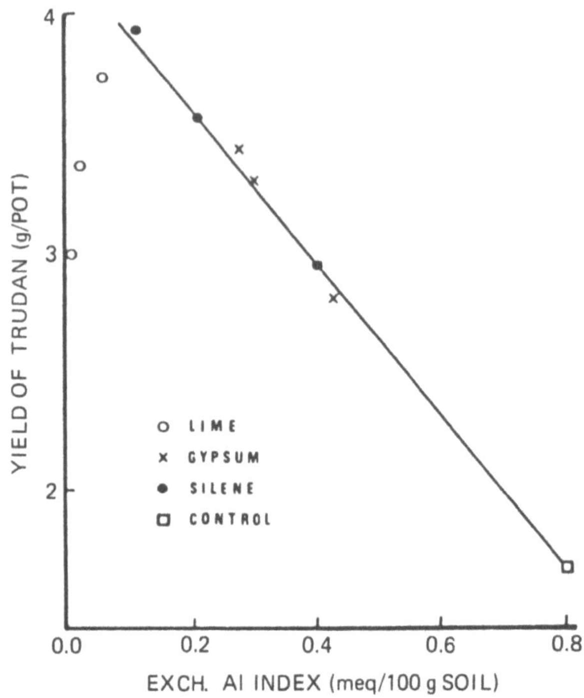

Figure 1. Relationship between exchangeable aluminum index (EAI) and yield of trudan on eight Natal Oxisols. 1 meq Al 100 g^{-1} = 1 $cmol_c$ kg^{-1} (Reeve and Sumner, 1970a).

The discussion following deals first in a general way with various measures of soil Al and Ca and their relationship to plant response. Thereafter problems involving Al and Ca in acid subsoils are defined more specifically, and possible methods of improving the rooting environment of these subsoils are discussed.

II. Soil Al and Plant Response

The importance of Al in soil acidity is widely recognized today and has become better understood in recent years (Jenny, 1961; Thomas and Hargrove, 1984). Physiological effects of Al on plants growing in acid soil have recently been reviewed (Foy, 1974, 1984). A major goal in soil and crop research has been to relate measures of soil acidity and soil Al to plant response which is required for an improved understanding of the phenomenon and as a basis for recommendations to growers.

A. Exchangeable Al and Al Saturation

Basing lime requirement on exchangeable Al has been found to give reasonable results for highly weathered soils (Evans and Kamprath, 1970; Kamprath, 1970;

Figure 2. Relationship between Al saturation of two Ultisols in Pennsylvania and relative corn grain yield. (Reproduced with permission from R.H. Fox, "Soil pH, aluminum saturation, and corn grain yield," *Soil Sci.* 127:330–334, copyright © by Williams & Wilkins, 1979.)

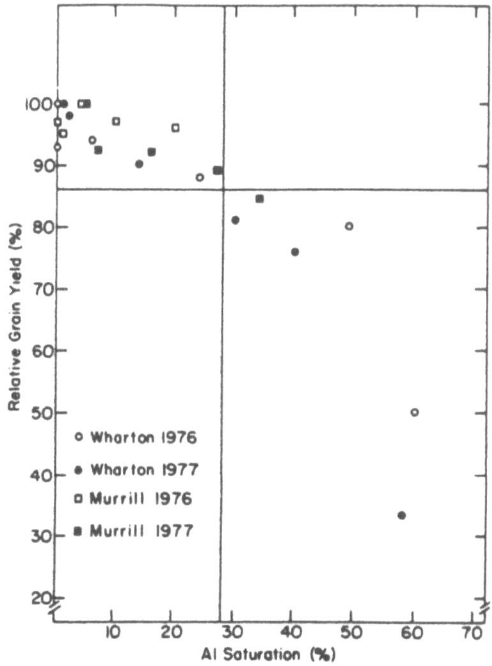

Reeve and Sumner, 1970b). Liming these soils to the often recommended pH values of 6.0 or 6.5 can require very large amounts of lime because of high buffer capacity, and it may also result in P or micronutrient deficiencies. Reeve and Sumner (1970a), working with eight Natal Oxisols with various ameliorants, were able to relate yield of Trudan (*Sorghum sudanense* L. X *Sorghum bicolor* L.) with a measure called exchangeable Al index (Figure 1). This index (EAI) was the amount of Al extracted in a single 2-minute shake with 0.2 *M* NH_4Cl at a 1:10 soil/solution ratio. Reeve and Sumner (1970b) found the average amount of lime necessary to achieve maximum Trudan growth and sufficient neutralization of exchangeable Al to be about one-sixth that required to raise the pH to 6.5. As Figure 1 shows, the higher rates of lime (open circles, EAI decreased as lime rate increased) used by Reeve and Sumner (1970a) resulted in substantial yield reductions.

Although exchangeable Al is a measure of reserve acidity and has been found to be related to root growth and yield (Moschler et al., 1960; Saigusa et al., 1980), it does not always relate well to plant response because of differences between soils in mineralogy, surface charge, organic matter, and other factors. A closer relationship with plant growth is often found with Al saturation percentage of the effective cation exchange capacity. Growth-limiting Al saturation levels have been found to vary from approximately 20 to 70%, depending on soil type and crop grown (Evans and Kamprath, 1970; Gonzales-Erico et al., 1979; Kamprath, 1970; Sartain and Kamprath, 1977). Fox (1979) found a critical Al saturation value of 28% for corn production on two Ultisols in Pennsylvania (Figure 2). The relationship between crop yield and Al saturation for two Ultisols and an

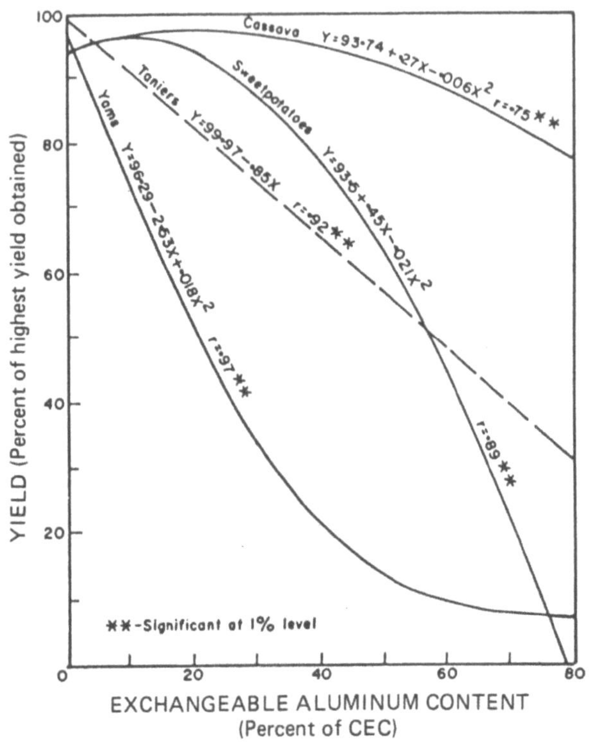

Figure 3. Relationship between Al saturation of two Ultisols and an Oxisol and yield of four tropical root crops in Puerto Rico (Abruna-Rodriguez et al., 1980).

Oxisol in Puerto Rico differed greatly for four tropical root crops (Abruna-Rodriguez et al., 1982) (Figure 3). Cochrane et al. (1980) suggested determining lime requirement by using an equation incorporating Al saturation for a given soil with the required Al saturation for the crop to be grown.

B. Soil Solution Al

The level of soil solution Al is a more direct measurement of the conditions to which a plant root responds than exchangeable Al or Al saturation. Various techniques and moisture regimes have been used to extract soil solution. Although time-consuming, such investigations have greatly increased our understanding of plant response to soil acidity, which has been further improved by studies of plant response to Al in nutrient solutions. Although there are obviously differences between growing a plant in nutrient solution and growing it in soil, nutrient and soil solution investigations can often be used together in attempting to explain plant response. Consequently, results from both types of study are considered in the discussion of solution Al that follows.

Soil solution Al^{3+} concentration has been found to be related to plant growth in numerous studies (Adams and Moore, 1983; Evans and Kamprath, 1970; Rag-

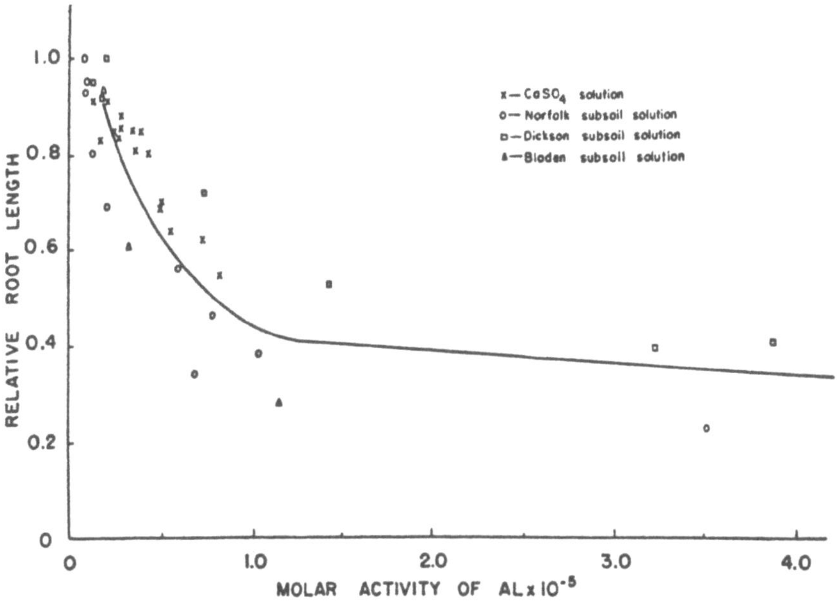

Figure 4. Effect of molar activity of Al in subsoil solutions in situ and in subsurface nutrient solutions on cotton primary root growth. (Reproduced with permission from F. Adams and Z.F. Lund, "Effect of chemical activity of soil solution aluminum on cotton root penetration on acid subsoils," *Soil Sci.* 101(3):193–198, copyright © by Williams & Wilkins, 1966.)

land and Coleman, 1959), but it is not as closely related to plant response as Al^{3+} activity. Some of the earlier studies showed strong relationships between plant growth and solution Al^{3+} activity corrected for ionic strength but not for Al complexes (Adams and Lund, 1966; Brenes and Pearson, 1973). Using this approach, Adams and Lund (1966) found that toxicity to cotton roots increased progressively as Al^{3+} activity exceeded a minimum of about 1.5 μM (pAl = 5.82) (Figure 4).

Complexation of Al with various ligands is known to alter the toxic effects of Al (Adams and Hathcock, 1984; Cameron et al., 1986; Hue et al., 1986). In a nutrient solution study with coffee (*Coffea arabica* L.) seedlings, Pavan and Bingham (1982) found that at an Al^{3+} activity of 12 μM (pAl = 4.92, calculated with the speciation program GEOCHEM; Sposito and Mattigod, 1980) root growth was reduced by 10% (Figure 5). In another study with coffee growth on Oxisols and Ultisols, Pavan et al. (1982) found that toxicity occurred above an Al^{3+} activity in a saturated soil extract of approximately 4.0 μM (pAl = 5.40) (Figure 6). In a field study with cotton (*Gossypium hirsutum* L.) on an Ultisol, McCray et al. (manuscript submitted for publication) found that yield was reduced when pAl in topsoil solution was less than 5.6 (calculated with GEOCHEM; Figure 7). These results show that even at very low Al^{3+} activities (pAl = 4.9–5.6), growth reduction is possible.

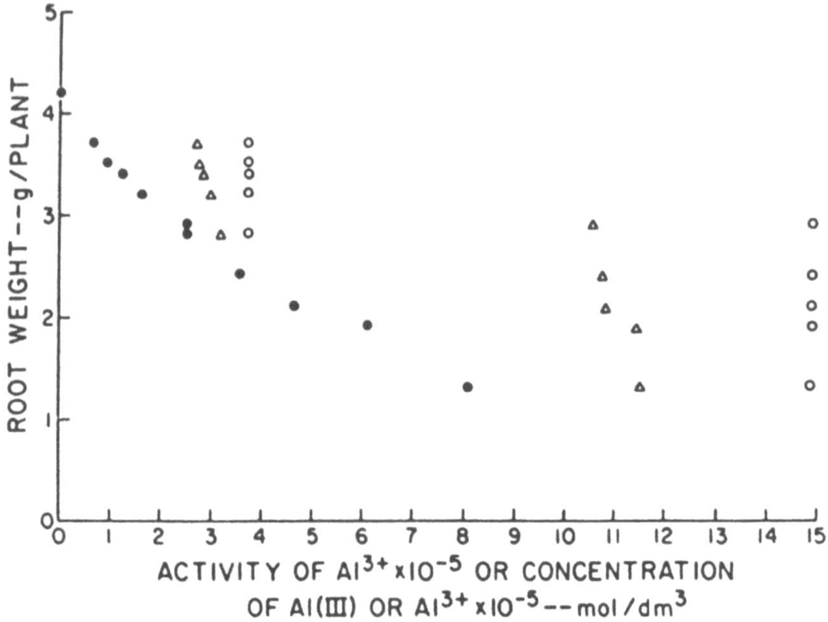

Figure 5. Coffee root weight in relation to activity of Al^{3+} (black circles) and concentrations of Al_t (open circles) and Al^{3+} (open triangles) in nutrient solutions (Pavan and Bingham, 1982).

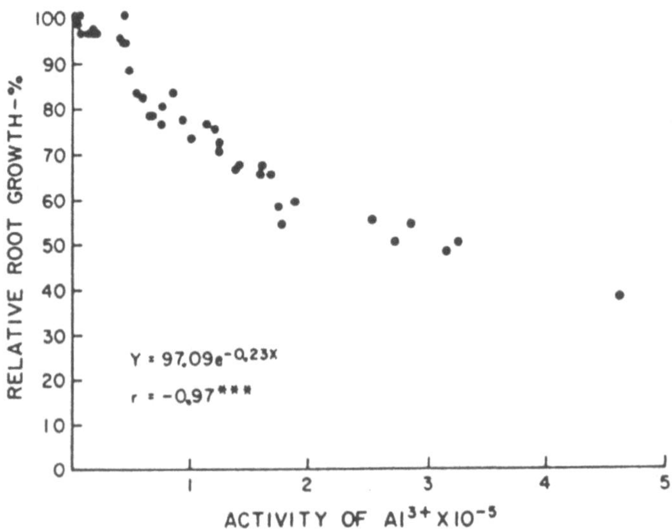

Figure 6. Relative root growth in relation to activity of Al^{3+} in soil solution (***: significant at the 0.001 level) (Pavan et al., 1982).

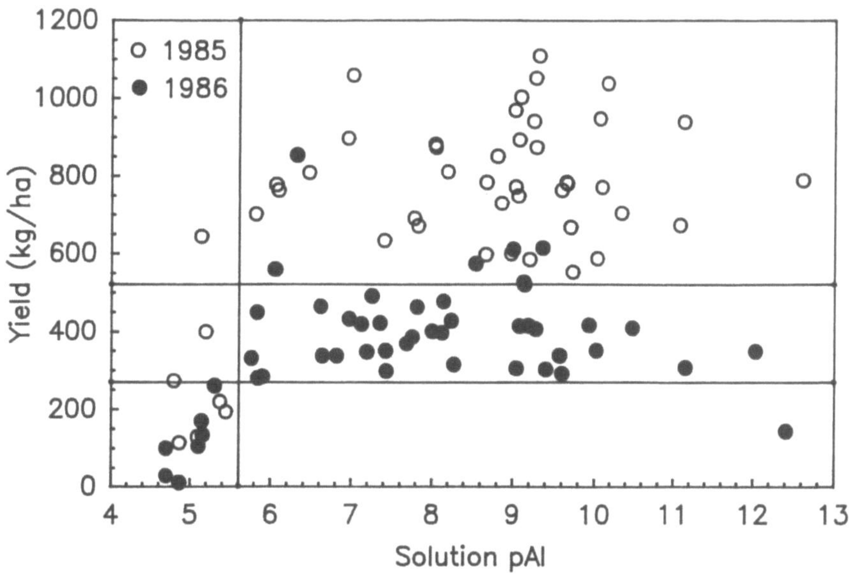

Figure 7. Relationship between soil solution Al^{3+} activity (0- to 0.2-m depth) and cotton lint yield in 1985 and 1986. Critical pAl (5.6) determined for each year's data with Cate and Nelson (1965) graphical technique (McCray et al., 1989).

C. Interaction of Al^{3+} with Inorganic and Organic Anions

The presence of inorganic and organic ligands must be considered in Al speciation to adequately reflect the properties of the particular solutions. Fluoride (F^-) and SO_4^{2-} are strong and weak Al^{3+}-complexing ions, respectively (Lindsay, 1979), but at high concentrations even SO_4^{2-} is an important ligand for Al, as will be discussed later in relation to gypsum application. As a result of hydrolysis, aluminum–hydroxy species can make up a substantial part of the total Al in solution. Equilibrium constants compiled by Lindsay (1979) indicate that other inorganic anions (other than phosphate) present in substantial quantities in soil solutions subtend only very low levels of Al complexes in solution, although they may indirectly influence Al^{3+} activity by increasing ionic strength.

The interaction between Al and P is important in terms of P adsorption on hydroxy Al sites (Parfitt, 1978) and precipitation of Al phosphates (Lindsay, 1979). Reduction of Al toxicity with added P has been reported in nutrient solution studies (Alva et al., 1986; Bartlett and Riego, 1972; Blamey et al., 1983) and with soil-grown plants (Awad et al., 1976; Bache and Crooke, 1981; Estrada and Cummings, 1968). Many highly weathered soils have a P availability problem in addition to high levels of soluble Al^{3+}, since these soils have such high P-fixing capacities (Fox and Searle, 1978). This limits the usefulness of using P to overcome Al toxicity, but it also points out the need to consider the relation between P availability and Al toxicity in a given soil.

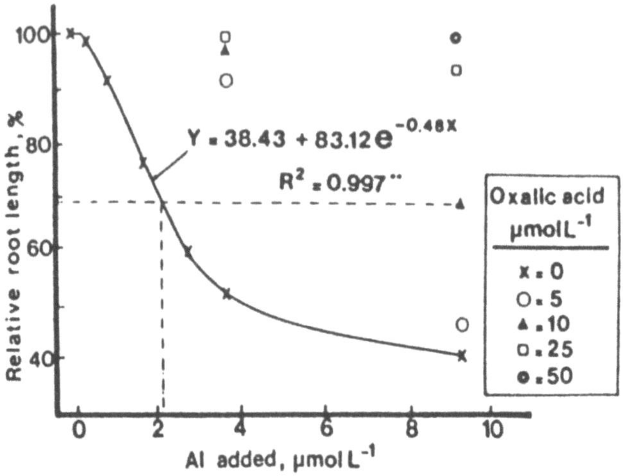

Figure 8. Relative root length as a function of initial concentrations of $AlCl_3$ and oxalic acid. The solid line presents Al effect in the absence of organic acid (standard curve for Al toxicity). The dashed line illustrates conversion of root growth in the presence of oxalic acid to equivalent monomeric Al concentration (Hue et al., 1986).

The amount of organic matter in a soil has been found to be a major factor in determining the toxicity of Al to plant growth (Hargrove and Thomas, 1981; Thomas and Hargrove, 1984). Evans and Kamprath (1970) found that as the organic matter content of soils increased, less Al^{3+} was present in the soil solution at a given pH. Bloom et al. (1979a) concluded that exchange of Al ions from organic matter exchange sites controls the relationship between pH and Al^{3+} activity in acid soils with a low amount of permanent charge CEC relative to the quantity of organic matter. Adams and Hathcock (1984) suggested that the primary difference in Al toxicity between soils examined in cultivated fields and woodlands was the amount of organically bound Al. Chemical methods were unable to distinguish organically bound Al from inorganic Al, but any amount of Al measured in cultivated fields was much more toxic than an equal amount measured in woodland soil. Hue et al. (1986) found wide variations in both kinds and concentrations of organic acids in soil solutions of different eluviated horizons. The organic acids most effective at alleviating Al toxicity were citric, oxalic, and tartaric acids in that decreasing order, while malic, malonic, and salicylic acids were less effective in that decreasing order. Figure 8 demonstrates the importance of these organic acids in reducing Al toxicity by showing the relationship between relative root length and solution Al at various oxalic acid concentrations (Hue et al., 1986).

Overliming of highly weathered soils can cause yield reductions as a result of increased solubilization and uptake of Al–organic matter complexes in the pH range of 5 to 7 (Farina et al., 1980; Hargrove, 1986). Although this effect has occurred only at near-neutral soil pH, which should be avoided in these soils

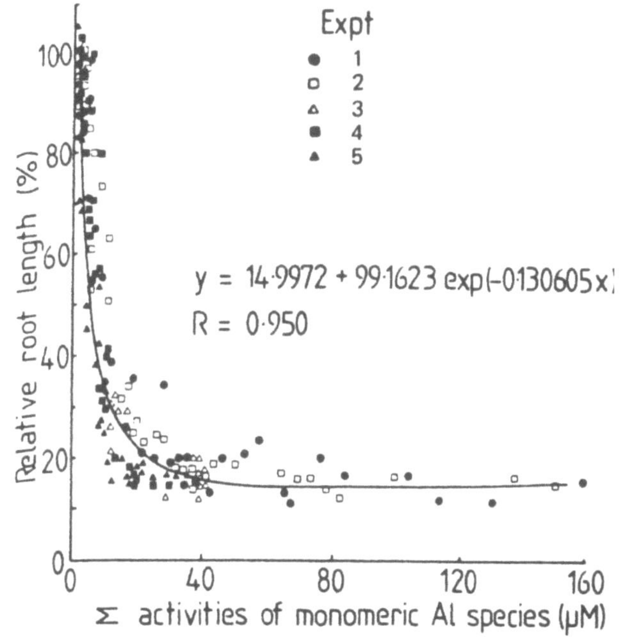

Figure 9. Rate of color development at 530 nm after the addition of aluminon buffer to three solutions containing an initial concentration of 50 μM Al. (Reproduced with permission from F.P.C. Blamey, D.G. Edwards, and C.J. Asher, "Effects of aluminum, OH:Al and P:Al molar ratios, and ionic strength on soybean root elongation in solution culture," *Soil Sci.* 136(4):197–207, copyright © by Williams & Wilkins, 1983.)

anyway, further research on these Al–organic matter complexes may be necessary. Since the major emphasis of this chapter is on subsoil acidity, organic matter is less important than in the case of topsoils.

Several researchers have suggested that mononuclear Al species other than Al^{3+} may be toxic to root growth. Blamey et al. (1983) found near-maximum Al toxicity to root growth at a total activity of Al monomers of 20 μM (Figure 9) but were not able to identify with certainty any specific Al monomer as being responsible for the toxicity. Indications are that Al complexed with F^- or SO_4^{2-} is not toxic to root growth (Cameron et al., 1986; Kinraide and Parker, 1987; Tanaka et al., 1987). These studies suggest that the major toxic Al species are Al^{3+}, $Al(OH)^{2+}$, and $Al(OH)_2^+$, with Al^{3+} being most toxic.

Polynuclear hydroxy-Al ions have been found to exist in dilute acid solutions of Al and orthophosphate (White et al., 1976) and in dilute salt extracts taken from acid soils (Bache and Sharp, 1976). This suggests that polynuclear Al species may exist in the soil solution, but significant quantities have not been demonstrated. Preferential adsorption of polynuclear Al by clay minerals and organic matter might prevent substantial concentrations from remaining in solution (Zelazny and Jardine, 1988). Parker et al. (1988) demonstrated that polynuclear hydroxy-Al was toxic to plants grown in nutrient solutions.

Table 1. Determination of lime requirement from pH and extractable soil Al

pH	Al µg2g soil	Limestone needed mg/g soil[a]
>6.6		0
	0–30	0.5
6.3–6.6	31–60	1.0
	>60	1.5
	0–30	1.0
5.6–6.2	31–60	2.0
	61–80	2.5
	>80	3.0
<5.6	0–70	2.5
<5.6	>70	*7.64logAl–11.6*

Source: Reprinted from R.J. Bartlett, "Reactive aluminum in the Vermont soil test," *Commun. Soil Sci. Plant Anal.* 13:497–506, 1982, by courtesy of Marcel Dekker, Inc.

[a] One unit represents 2 metric tons/hectare or 1 English ton/acre.

Various complexing agents have been used to attempt to distinguish between mononuclear and polynuclear Al species in solution (Barnhisel and Bertsch, 1982). Three commonly used spectrophotometric methods for determination of phytotoxic Al use ferron (Davenport, 1949), 8-hydroxyquinoline (James et al., 1983; Okura et al., 1962) or aluminon (Blamey et al., 1983; Hsu, 1963) as color-developing reagents. The complexity of the systems being studied has ensured that the use of these complexing agents to estimate phytotoxic Al will be difficult. A study by Bertsch et al. (1986) using wet chemical and Al-27 NMR methods to speciate hydroxy–Al solutions indicated that characterizations of these solutions without direct measurement can lead to incorrect estimations of mononuclear and polynuclear Al present. Also, the concentrations of other ions in solution can have a substantial effect on the usefulness of a particular colorimetric technique. Aluminum measured by the 8-hydroxyquinoline (15-s reaction) method has been found to be a better predictor of Al phytotoxicity than that measured by the aluminon (30-minute reaction) method in solutions with varying concentrations of F^- (Alva et al., 1988; Noble et al., 1988b). Alva et al. (1989) found that decreasing the reaction time of the aluminon reagent to 15 seconds appeared to exclude Al complexed with F^-. In that same study, 8-hydroxyquinoline, aluminon, and ferron failed to exclude Al complexed with SO_4^{2-}, regardless of reaction time. Jardine and Zelazny (1987,a,b) found that the ability of ferron to distinguish between mononuclear and polynuclear Al in solutions with various organic or inorganic anions often depended on specific anion/Al ratios. Using soil solutions in which concentrations of organic acids and dissolved organic C were low, Wright et al. (1987) found that both 8-hydroxyquinoline and aluminon gave reasonable estimates of phytotoxic Al. Hodges (1987) found the 8-hydroxyquinoline and ferron methods to have only limited usefulness for estimating phytotoxic Al in natural samples high in dissolved organic carbon, since the reagents degrade a portion of the Al–organic matter

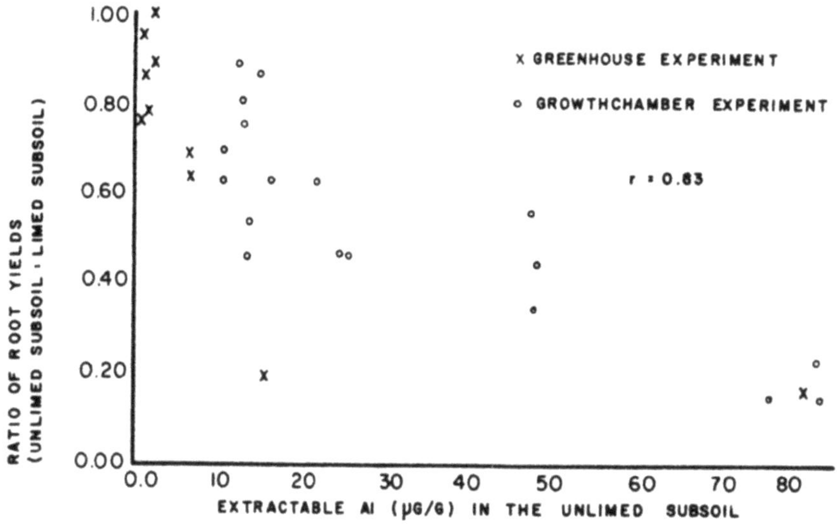

Figure 10. Relationship of alfalfa root yields to 0.02 *M* $CaCl_2$ extractable soil Al (McKenzie and Nyborg, 1984).

complexes. In that same study, as an alternative to colorimetric methods, Hodges (1987) suggested that the fluoride-sensitive electrode holds promise for the quantification of Al^{3+} activity in solutions in which Al-bound F^- constitutes a major portion of the total F^- and pH is less than 5.5.

D. Other Extraction Methods

Categories other than *M* KCl exchangeable and soil solution Al have been assayed in attempts to relate soil Al with plant response. Chloride salts other than KCl have been evaluated as extractants of soil Al. In particular, $LaCl_3$ appears to hold promise as an extractant for determining lime requirement, since it removes organically bound Al which KCl does not (Bloom et al., 1979b; Hargrove and Thomas, 1984; Oates and Kamprath, 1983a). Working with soils ranging in organic matter content from 0.3 to 58.3%, Oates and Kamprath (1983b) found that the lime rate predicted by Al removed with $LaCl_3$ reduced the residual Al to <15% Al saturation for all soils, while the lime rate predicted by $CuCl_2$ was in excess of that needed. The Vermont soil test for "reactive Al" uses an extraction with pH 4.8 NH_4OAc (1.25 *M* acetate; Bartlett, 1982). In this test designed for acid soils of the northern United States, extracted Al is used in conjunction with pH in 0.01 *M* $CaCl_2$ to calculate lime requirement (Table 1). Dilute $CaCl_2$ has been used as a soil extractant for some time because it closely approximates the ionic strength of moving soil solutions. McKenzie and Nyborg (1984) found that Al extracted with 0.02 *M* $CaCl_2$ was a more accurate indicator of root yield impairment in acid subsoils than soil pH or exchangeable Al (Figure 10). A limitation of dilute $CaCl_2$ as an extractant for Al is that it has been found to be a poor predictor of Al toxicity in soils containing soluble F^- (Moore and Ritchie, 1988).

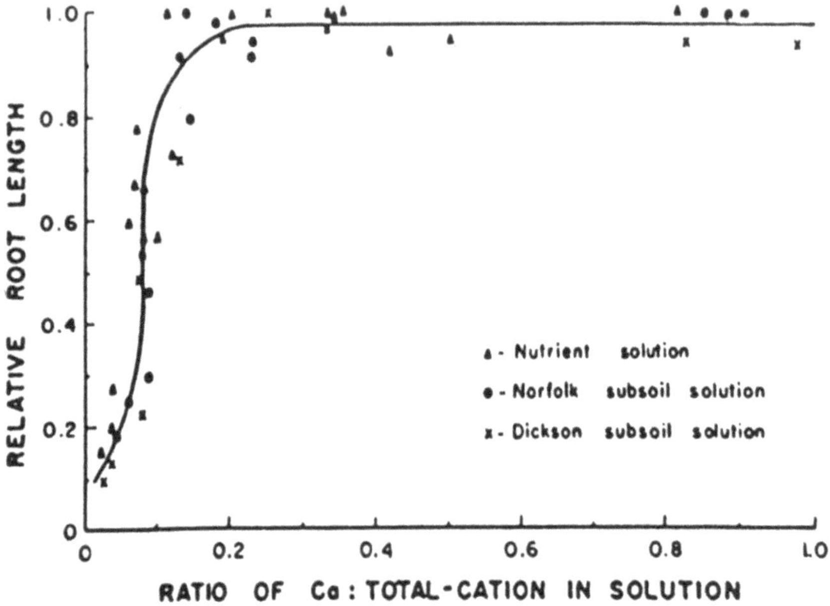

Figure 11. Effect of Ca/total cation ratio in solution on the penetration of subsurface medium by primary roots of cotton seedlings (Howard and Adams, 1965).

III. Ca Deficiency and the Ca/Al Relationship

Because Ca is transported very slowly in the phloem, and is necessary for cell elongation and cell membrane function, it must be continually supplied to the root meristem from the growth mechanism (Gauch, 1972; Emanuelsson, 1984). Calcium requirements for root growth have been found to be very low when other cations are in balance (Burstrom, 1952; Jones and Lunt, 1967; Lund, 1970; Tanaka and Woods, 1972). Kirkby (1979) concluded that in most soils, nutritional problems related to Ca are more likely to be caused by impaired translocation of or antagonism in the uptake of NH_4^+, K^+, Mg^{2+}, or Al^{3+} ions than by a simple Ca deficiency. The antagonistic influence of other cations on Ca nutrition was demonstrated in nutrient solutions and soils by Howard and Adams (1965; Figure 11) and Lund (1970; Figure 12).

From a practical point of view, higher soil Ca levels than might be necessary in a nutrient solution are usually required because the ideal situation is seldom found in a real soil. Howard and Adams (1965) found the lower limit of Ca^{2+} concentration needed for optimum root growth in Norfolk and Dickson subsoil solutions to range from 200 to 500 μM (0.4 to 1.0 $cmol_c\ kg^{-1}$; Figure 13). In a study of subsoils from the Coastal Plain of the southeastern United States, Adams and Moore (1983) found that low soil Ca limited root growth in soil horizons with soil solution Ca^{2+} activity $\leq 270\ \mu M$ (pCa $\geq$ 3.57) and Ca^{2+} saturation $< 17\%$ of the

Figure 12. Effect of ratio of molar activity of Ca to activity of cations on soybean root growth in subsurface nutrient solutions (Lund, 1970).

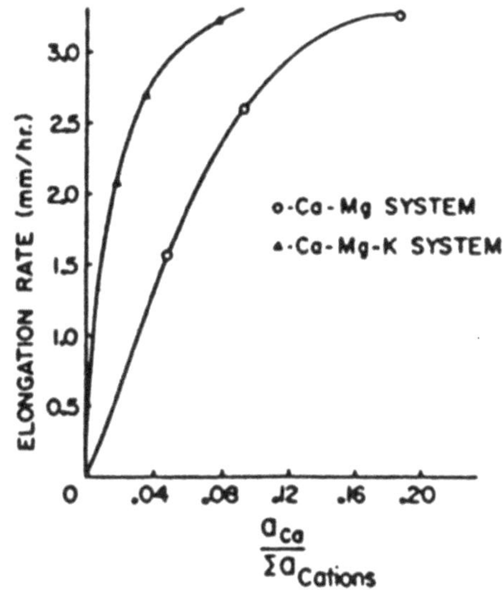

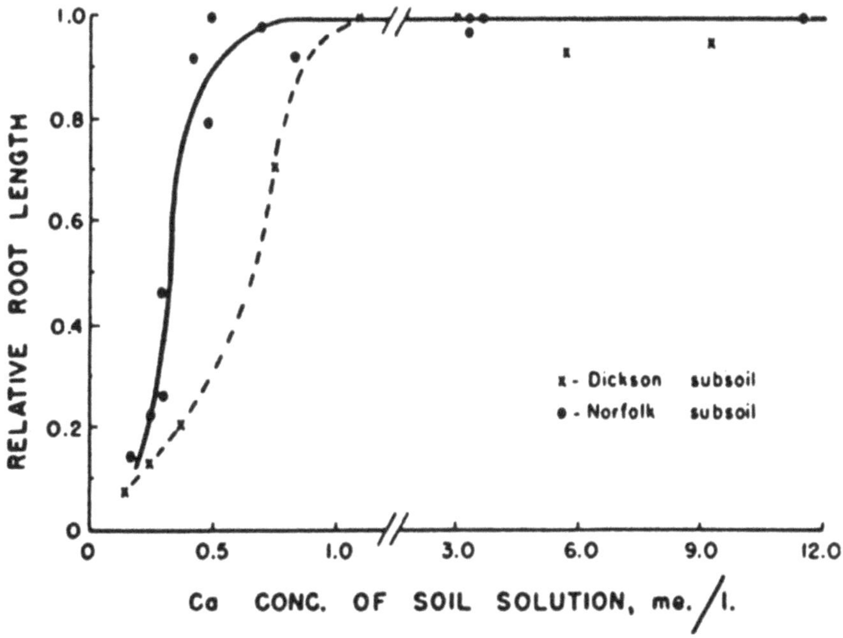

Figure 13. Effect of Ca concentration of subsoil solutions in situ on penetration of subsoil by primary roots of cotton seedlings. 1 meq Ca L^{-1} = 0.5 m*M* Ca (Howard and Adams, 1965).

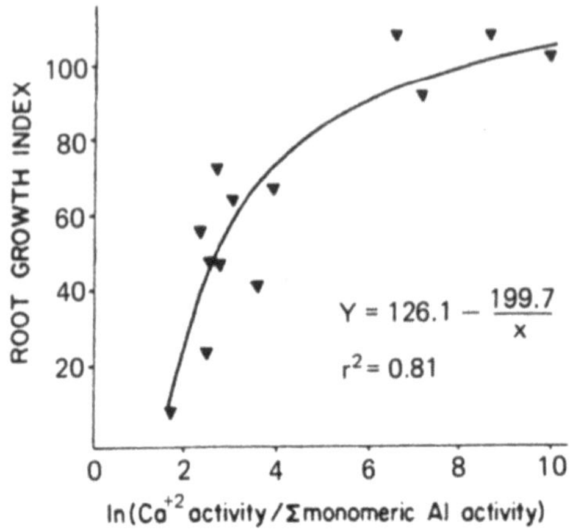

Figure 14. Subterranean clover root growth index as a function of ln ($a_{Ca^{2+}}/a_{monomeric\ Al}$) in soil solutions of unlimed soils. (Reproduced with permission from R.J. Wright and S.F. Wright, "Effects of aluminum and calcium on the growth of subterranean clover in Appalachian soils," *Soil Sci.* 143:341–348, copyright © by Williams & Wilkins, 1987.)

ECEC. Antagonistic cations were involved in the Ca response in each of these studies and should be expected to some degree in most soils.

Calcium and Al in the soil or nutrient solution have been shown to have an antagonistic relationship in numerous studies. Antagonistic effects between these two elements include decreased susceptibility to Al toxicity at increased solution Ca^{2+} levels, and reduced uptake and translocation of Ca as solution Al^{3+} concentration is increased (Alva et al., 1986; Clarkson and Sanderson, 1971; Evans and Kamprath, 1970; Lance and Pearson, 1969; Lee and Pritchard, 1984; Lund, 1970).

Because of the antagonistic effects between Al and Ca, it has been postulated that the calcium aluminum activity ratio [0.5 log ($a_{Ca^{2+}}$) − 0.33 log ($a_{Al^{3+}}$)] may provide a meaningful index by which rooting performance of plants growing in acid soils may be anticipated (Sumner et al., 1986). Wright and Wright (1987) used the ratio of Ca^{2+} activity and sum of predicted activities of monomeric Al species (Al^{3+} + hydroxy-Al + $AlSO_4^+$) for predicting the root growth of subterranean clover (*Trifolium subterraneum* L.; Figure 14). Noble et al. (1988a,c), working with nutrient solutions, related root length of soybean (*Glycine max* (L.) Merr.) to a function called the calcium aluminum balance (CAB) [2 log ($a_{Ca^{2+}}$)] − [3 log ($a_{Al^{3+}}$) + 2 log ($a_{Al(OH)^{2+}}$) + log($a_{Al(OH)_2^+}$)] (Figure 15), which was better related to root growth than Al^{3+} activity or the molar ratio of activities Al^{3+}: Ca^{2+}. Studies by Noble et al. (1988c) and Noble and Sumner (1988) also demonstrated the phytotoxicity of $Al(OH)^{2+}$ and $Al(OH)_2^+$, and the influence of ionic valence on degree of phytotoxicity. Regardless of specific functions used, it is evident that the relative levels of Ca and Al in acid soils must be considered in addition to actual levels of each.

IV. Ca and Al in Acid Subsoils

A. Defining the Problem

Acid soils have developed under humid conditions where weathering processes have resulted in Al^{3+} occupying a large proportion of the exchange sites. This

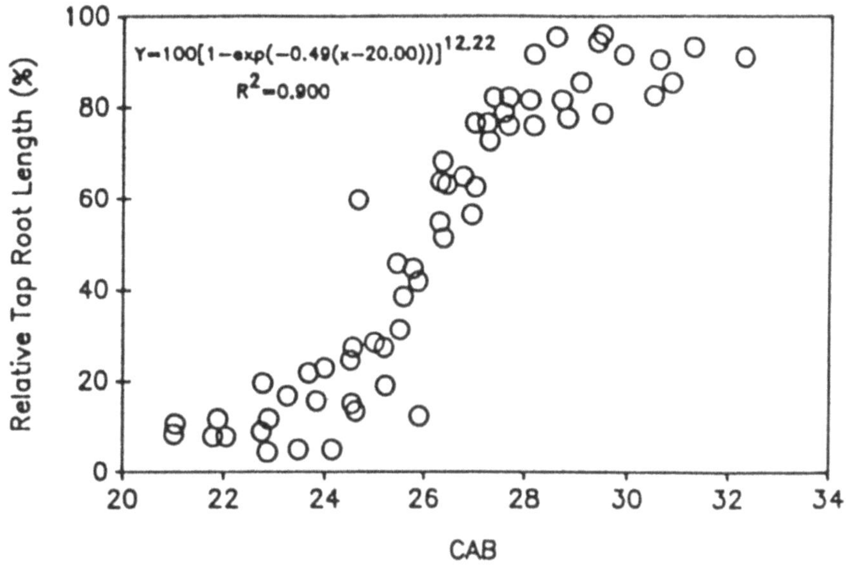

Figure 15. Relationship between relative tap root length of soybeans grown in dilute nutrient solutions and CAB. Nonlinear analysis using the marquardt iterative method (Noble et al., 1988a).

acidification becomes a serious problem below the normal plow depth. Since acid subsoils have been reported in colder climates such as in Canada (McKenzie and Nyborg, 1984), subsoil acidity is certainly not limited to the warmer regions of the world. In the case of acid subsoils studied by McKenzie and Nyborg (1984), however, toxic amounts of Al were attributed to the parent material (marine shales) rather than to the degree of pedogenic weathering. The processes that have formed acid subsoils are very important in determining the specific nature of problems and appropriate ameliorative treatments.

Where acid subsoils have developed through advanced pedogenic weathering processes, weathering intensity is an important factor. Characteristics of the two soils shown in Table 2 (Pavan et al., 1984; Sumner et al., 1986) demonstrate the progression of weathering resulting in increased soil solution Al^{3+}, which represents a barrier to root growth. The Georgia soil occurs on the Piedmont of the southeastern United States and, although weathered to a high degree as indicated by the high Al saturation, has much lower soil solution Al^{3+} concentrations throughout the soil profile than the much more highly weathered Brazilian Oxisol from the tropics. Other studies have found soil solution Al levels to be relatively low in subsoils of the southeastern United States. With the exception of an eluviated horizon, Adams and Moore (1983) found total soil solution Al concentrations to be no greater than 17 μM in six Coastal Plain subsoils of the southeastern United States. Pavan et al. (1982) found total solution Al concentrations of 51.3 and 58.9 μM in the upper 0.30 m of an Oxisol and Ultisol, respectively, from Brazil. The limited amount of available soil solution data prevents

Table 2. Comparison of selected characteristics of acid soils from the southeastern U.S. and Brazil

Soil depth (m)	Solution concentrations[a] (μM) Al_T	Al^{3+}	Ca^{2+}	H_2O pH	Exch. Al ($cmol_c\ kg^{-1}$)	Exch. Ca ($cmol_c\ kg^{-1}$)	Al Sat. (%)
			Georgia Ultisol (Appling series)[b]				
0–0.15	0.0	0.0	1680	6.2	0.00	3.75	0
0.15–0.30	13.0	8.4	410	5.0	1.40	0.95	45
0.30–0.45	12.0	8.4	400	4.7	1.52	0.75	50
0.45–0.60	11.0	7.8	250	4.7	1.70	0.60	63
0.60–0.75	10.0	6.8	160	4.9	1.86	0.30	78
0.75–0.90	9.0	5.8	70	4.8	1.98	0.35	79
0.90–1.05	10.0	6.6	40	4.9	2.24	0.24	85
			Brazilian Oxisol[c]				
0–0.05	0.0	0.0	4902	–	–	–	–
0.05–0.10	0.0	0.0	3538	–	–	–	–
0.10–0.20	15.0	7.5	410	–	–	–	–
0.20–0.40	23.0	16.1	40	4.4	1.65	1.66	41
0.40–0.60	31.0	23.6	50	4.3	1.82	1.30	49
0.60–0.80	50.0	40.0	60	4.2	1.88	1.12	54
0.80–1.00	53.0	42.4	60	4.1	2.05	0.80	65

[a] Saturated paste method. Al_T = Total solution Al.

[b] This soil (clayey, kaolinitic thermic Typic Hapludult) was located in Watkinsville, Ga., and had received 3 Mg fine calcitic lime ha^{-1} mixed in the topsoil 2.08 years before sampling for these properties (Sumner et al., 1986).

[c] This soil taken as an undisturbed column from Londrina, Brazil, had received sufficient $CaCO_3$ mixed in the upper 0.02 m to theoretically neutralize 150% of exchangeable Al per column, and was subsequently leached for 6 months. Soil pH and exchangeable cations were not available for this treatment, so values from untreated columns were included for depths below 0.20 m (Pavan et al., 1984).

comparisons of solution of Al levels between large numbers of soil types. Available data indicate that generally, higher concentrations of solution Al^{3+} occur in the tropics, and so ameliorating these acid soils can be particularly difficult. The lower levels of solution Al found in regions of less intense weathering can be very important as well, since small amounts of Al^{3+} can be toxic.

Various acid subsoils developed through advanced pedogenic weathering are similar in that they have low levels of exchangeable and solution Ca^{2+} (Adams and Moore, 1983; Farina and Channon, 1988b; Pavan et al., 1984; Sumner et al., 1986). Even if there was not an Al problem in these subsoils, Ca insufficiency would often exist, as in the examples given in Table 2 (Adams and Moore, 1983; Howard and Adams, 1965). Since the relative levels of Ca^{2+} and Al^{3+} are important in root growth, it follows that the most difficult subsoils to ameliorate would be those with high solution Al^{3+} activity in addition to low Ca^{2+} activity. The unamended solution Al^{3+} activity would be an important factor in determining the potential benefit from an amendment.

In situations in which acid subsoils have developed because of the presence of a particular kind of parent material rather than through advanced pedogenic weathering, toxic amounts of Al^{3+} are likely to be the major problem. McKenzie and Nyborg (1984) found barley (*Hordeum vulgare* L.) and alfalfa (*Medicago sativa* L.) top growth and root penetration in such subsoils to be well related to 0.02 *M* $CaCl_2$–extractable Al. Soil solution Ca^{2+} activity might be less useful as an indicator of a problem in these soils since they have not been weathered as intensely.

If the horizons in question were nearer the surface, the obvious answer to the problem would be to incorporate lime to increase and decrease solution Ca^{2+} and Al^{3+}, respectively. The difficulty and expense of deep lime incorporation prevents it from being a more viable alternative, although deep lime placement is discussed later as a possible treatment. Thorough mixing of lime into the target soil is necessary because of slow downward movement due to its low solubility, neutralization of OH^- and HCO_3^- ions by H^+ and Al^{3+} ions in surface horizons, and Ca^{2+} adsorption by pH-dependent sites generated as pH increases. The inability of surface-incorporated lime to ameliorate deeper soil horizons has been demonstrated in various studies (Khalid and Silva, 1980; Pavan et al., 1982; Reeve and Sumner, 1972; Ritchey et al., 1980).

The presence of subsoil compacted layers often complicates the chemical and nutritional problems in acid subsoils. Since root growth in a particular soil horizon is inversely related to soil strength (Gerard et al., 1972; Taylor et al., 1966), physical profile modification might have to be considered as a treatment in addition to chemical amelioration.

There is evidence that structural problems may exist in many acid subsoils as a result of the very low ionic strength of the soil solution (Sumner et al., 1990). This may actually be less of a problem in subsoils that have very high levels of solution Al^{3+}, as can occur in the tropics and subtropics, because of the flocculating effect of Al^{3+}. The high flocculating power of the Al^{3+} ion stems from its small hydrated radius and has been demonstrated in terms of its interaction with organic matter (Khan, 1969; Wright and Schnitzer, 1963). Of course, at high solution Al^{3+} activity, the major problem is of a chemical/nutritional nature.

B. Means of Improvement

Yield responses have been obtained to deep lime placement (Brennan and Axley, 1983). Farina and Channon (1988a) successfully used a deep limer (Figure 16) and a modified subsoiler (Figure 17) to mix lime and limed topsoil, respectively, into the subsoil trench. Anderson and Hendrick (1983) used a subsoil lime injector (Figure 18) to apply a lime slurry into the subsoil. A limitation of these and similar methods of liming the subsoil is that a relatively low volume of soil is ameliorated for the high operational costs and power requirements. Of course, in situations in which high subsoil strength and acid subsoil are both limiting factors, the combination of subsoiling and deep lime placement becomes more attractive.

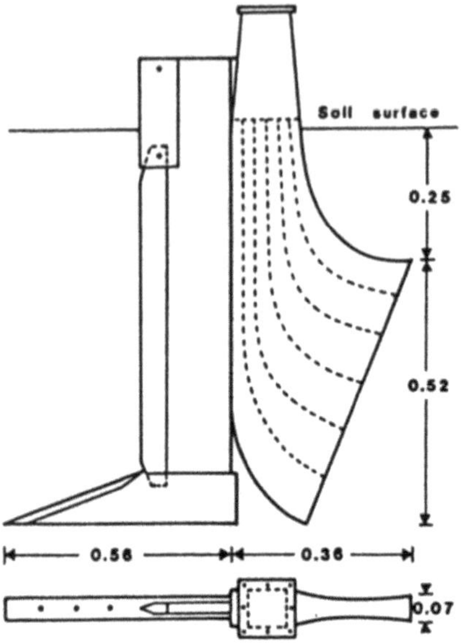

Figure 16. Sideview and plan of deep limer (specifications in meters) (Farina and Channon, 1988a).

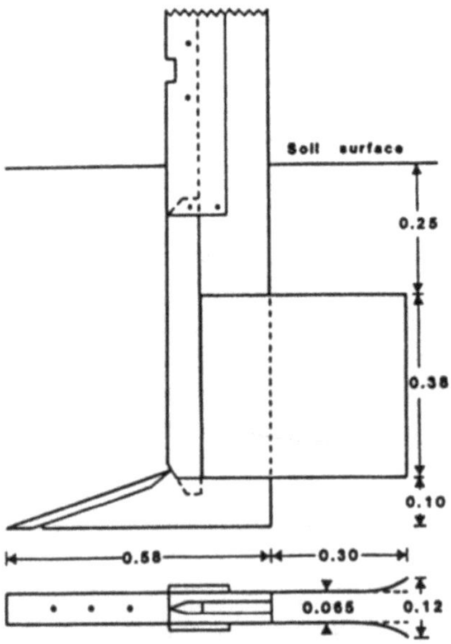

Figure 17. Sideview and plan of modified subsoiler (specifications in meters) (Farina and Channon, 1988a).

The slow movement of lime through the soil and the expense and difficulty of mixing lime into the subsoil have led researchers to examine surface-applied amendments, such as gypsum, in the amelioration of subsoil acidity. Since gypsum ($CaSO_4 \cdot 2H_2O$) does not alter soil pH appreciably, pH-dependent charge remains unchanged, enabling the gypsum to move through the soil much more

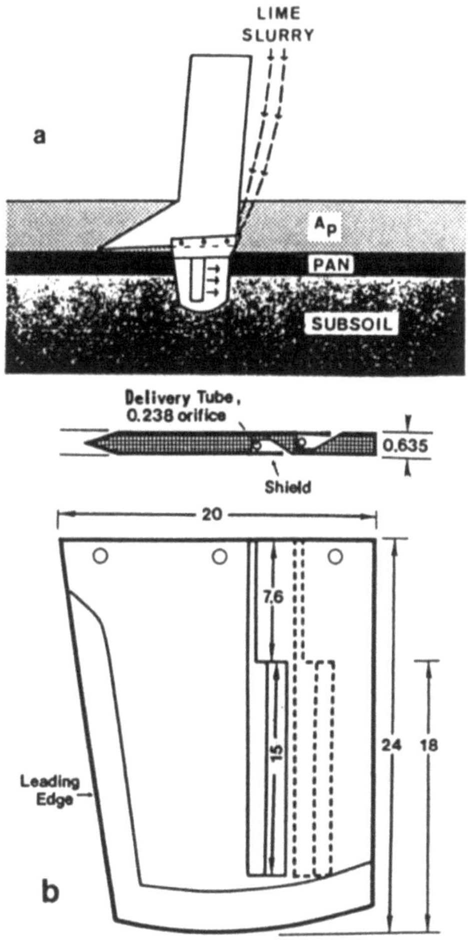

Figure 18. Overall relationship between the chisel, slitter, and soil layers (a) and close-up detail of the slitter (b) (specifications in centimeters) (Anderson and Hendrick, 1983).

readily than lime. Sumner (1970) and Reeve and Sumner (1972) were the first to demonstrate the beneficial effects of gypsum on subsoil acidity. They found that surface-incorporated gypsum moved through a simulated profile of a Natal Oxisol and reduced the level of exchangeable Al in the subsoil more effectively than surface-incorporated lime (Figure 19). Ritchey et al. (1980) found that $CaSO_4$ and $CaCl_2$ moved deeper than $CaCO_3$ into disturbed columns of an Oxisol from Brazil (Figure 20). The rate of $CaCl_2$ movement through the column was faster than $CaSO_4$ because of the higher reactivity of SO_4^{2-} with soil surfaces and the higher solubility of $CaCl_2$. However, $CaCl_2$ should not be considered as a potential ameliorant for subsoil acidity because of potential salt toxicity to crops.

Increased yields of crops grown on acid soils with surface-incorporated gypsum have been attributed to increased exchangeable Ca, decreased exchangeable Al, and increased root length density in the subsoil (Farina and Channon, 1988b;

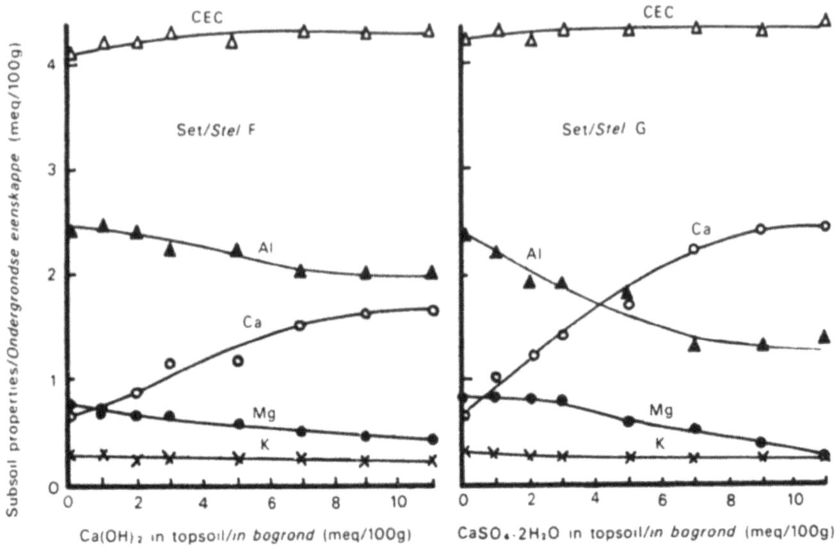

Figure 19. CEC and exchangeable cations in Clovelly subsoil (15–30 cm) after leaching with effluent from ameliorated topsoil. 1 meq 100 g^{-1} = 1 $cmol_c$ kg^{-1} (Reeve and Sumner, 1972).

Hammel et al., 1985; Sumner et al., 1986). The lack of yield response to gypsum application in other studies (Black and Cameron, 1984; Simpson et al., 1979; Wright et al., 1985) has shown the importance of understanding the reasons for observed responses so that proper recommendations regarding the use of gypsum can be made.

Since the activities of Ca^{2+}, Al^{3+}, other species of Ca and Al, and the balance of these ions in the soil solution directly affect root growth, any possible response to gypsum in an acid subsoil must ultimately be discussed in terms of these parameters. Greatly increased subsoil solution Ca^{2+} activity after gypsum application (Table 3; McCray et al., 1989; Sumner et al., 1986) certainly plays a major role in improved root growth and overall plant response. Solution Al^{3+} concentration may actually be increased with gypsum application, but due to increased ionic strength and formation of the $AlSO_4^+$ ion pair (Pavan et al., 1982; Singh, 1982), Al^{3+} activity is often reduced.

Surface charge is important in solution Al^{3+} activity changes with gypsum application. The presence of a substantial amount of positive charge may prevent larger amounts of Al^{3+} from being released as Ca^{2+} activity and ionic strength are increased. This would be related to pH changes as ionic strength increases and to possible SO_4^{2-} for OH^- ligand exchange as SO_4^{2-} concentration increases ("self-liming effect"–Reeve and Sumner, 1970a). Reduced Al^{3+} activity would be a less probable result in a soil with a very low pH_0 [point on the pH scale where there is an equal number of protonated and deprotonated sites on the variable

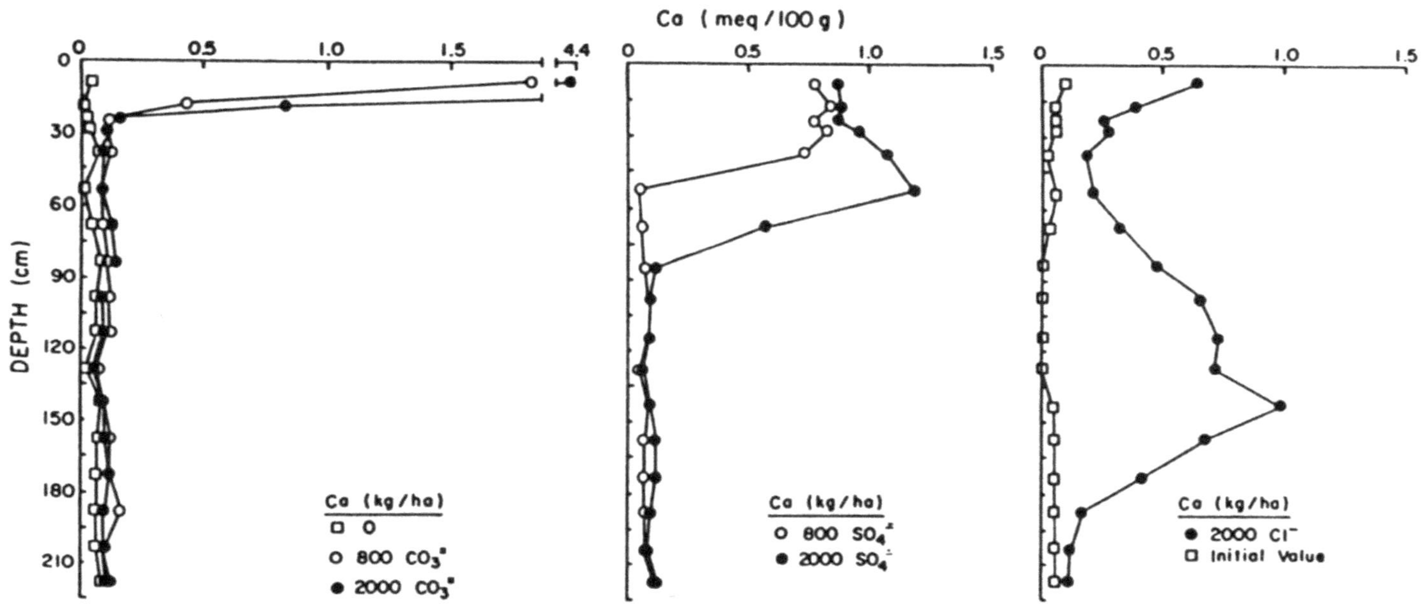

Figure 20. Results of a column experiment showing the effects of various anions on the distribution of Ca after leaching with the equivalent of 1200 mm of rainfall in a reconstructed virgin Dark Red Latosol profile 0 to 135 cm. Calcium as carbonate, sulfate, or chloride was added to the 0- to 15-cm layer and incubated 3 weeks before leaching began. 1 meq 100 g^{-1} = 1 $cmol_c$ kg^{-1} (Ritchey et al., 1980).

Table 3. Effect of gypsum application on soil solution Ca^{2+} activity in two field studies

Sumner et al. (1986)[a]			McCray et al. (1989)[b]		
	Ca^{2+} Activity (μM)			Ca^{2+} Activity (μM)	
Soil depth (m)	Control	Gypsum	Soil depth (m)	Control	Gypsum
0–0.15	1120	3630	0–0.20	1778	4169
0.15–0.30	344	2537	0.20–0.40	603	2344
0.30–0.45	174	1486	0.40–0.60	339	933
0.45–0.60	114	1123	0.60–0.80	195	389
0.60–0.75	72	741			
0.75–0.90	37	606			
0.90–1.05	28	408			

[a] Study conducted on an Appling coarse sandy loam (clayey, kaolinitic, thermic Typic Hapludult) with lime (3 Mg ha^{-1} to control and gypsum plots) and gypsum (10 Mg ha^{-1}) mixed into the topsoil 2.08 years before these samples were taken. Saturated paste method was used.

[b] Study conducted on an Appling loamy coarse sand and a Cecil coarse sandy loam (clayey, kaolinitic, thermic Typic Hapludults) with no lime applied in these treatments. Gypsum (10 Mg ha^{-1}) was mixed into the topsoil 1.33 years before these samples were taken. The soil solution was extracted with centrifuge method at -0.01 MPa matric potential.

charge surfaces in the soil (Uehara and Gillman, 1981)]. This would also mean that a favorable response to gypsum would be more likely where toxic levels of Al exist in a subsoil than in a topsoil, since increased organic matter present in topsoil would result in a lower pH_0 (Uehara and Gillman, 1981).

Many studies have shown that gypsum can decrease exchangeable Al as determined by extraction with an unbuffered salt solution (Farina and Channon, 1988b; Reeve and Sumner, 1972; Sumner et al., 1986). The mechanism of this change has been the subject of some debate. The importance of SO_4^{2-} in this response was demonstrated by Chao et al. (1962, 1963) by showing that exchangeable Al is correlated with the amount of SO_4^{2-} adsorbed by a soil. Reeve and Sumner (1970a) proposed that the reduction in exchangeable Al after gypsum application was the result of a "self-liming" effect resulting from SO_4^{2-}-induced Al polymerization by ligand exchange for OH^- groups (Chang and Thomas, 1963). This effect is the result of an adsorption-type reaction (Hue et al., 1985; Rajan, 1978) such as that shown in Figure 21. Singh (1982) reported an increase in pH in the range of 0.20 to 0.28 units when pH of acid soils was measured in 0.01 M $CaSO_4$ compared with 0.01 M $CaCl_2$. Some exchange of SO_4^{2-} for OH^- may take place when gypsum is added without a detectable increase in soil pH (Sumner et al., 1986). Measured pH in this situation will be the result of H^+ (Al^{3+}) exchanged by the increased salt concentration in solution and OH^- released by SO_4^{2-}–soil reactions (Hue et al., 1985). Farina and Channon (1988b) found an increase in soil-water pH with gypsum application in the zone of maximum sulfate sorption/precipitation. In general, however, subsoil pH changes have been inconsistent and are not a reliable indication of a favorable gypsum response.

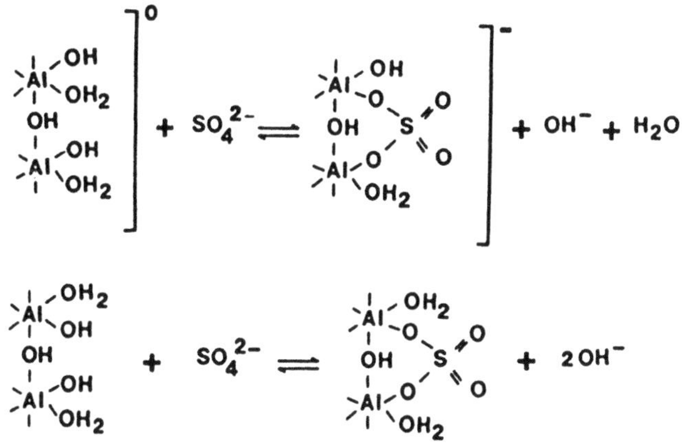

Figure 21. Adsorption type reactions of sulfate with hydroxy–Al surfaces (Hue et al., 1985).

Precipitation reactions (Adams and Rawajfih, 1977; Hue et al., 1985; Nordstrom, 1982) such as the following have also been proposed as a mechanism for reduction in exchangeable Al with gypsum application.

$$Al(OH)_3 + SO_4^{2-} \rightarrow Al(OH)SO_4 + 2OH^- \quad [1]$$

$$4Al(OH)_3 + SO_4^{2-} \rightarrow Al_4(OH)_{10}SO_4 + 2OH^- \quad [2]$$

The formation of Al hydroxy sulfate minerals may be important in situations in which pH is below 5.0, solution Al is present, and SO_4^{2-} has been added. Nordstrom (1982) suggests that with approximately m*M* SO_4^{2-} in solution, alunite ($KAl_3[OH]_6[SO_4]_2$) will be more stable than kaolinite below about pH 4.5 and jurbanite ($AlOHSO_4$) will be the most stable Al-containing solid only at pH values well below 4.0. Sposito (1985) suggested that gypsum treatment of a sesquioxidic soil might result in the formation of metastable basaluminite ($Al_4[OH]_{10}SO_4$), which would then slowly be converted to a more stable phase. In a column study with a Cecil soil, Evans and Zelazny (1987) attributed decreases in exchangeable Al and SO_4^{2-} in the Btl horizon to formation of $AlOHSO_4$ and basaluminite.

Parenthetically, in addition to the precipitation reactions involving SO_4^{2-}, precipitation reactions involving Si have been suggested by studies in which an increase in soil solution Si concentration was found after gypsum application (Gillman and Sumner, 1987; Lemus-Grob, 1985). The speciation program GEOCHEM (Sposito and Mattigod, 1980) suggests that the Si might precipitate with the soluble Al to form an alumino-silicate, but the stability constants used are questionable. The role of Si in changes in solution Al^{3+} with gypsum application is continuing to be investigated.

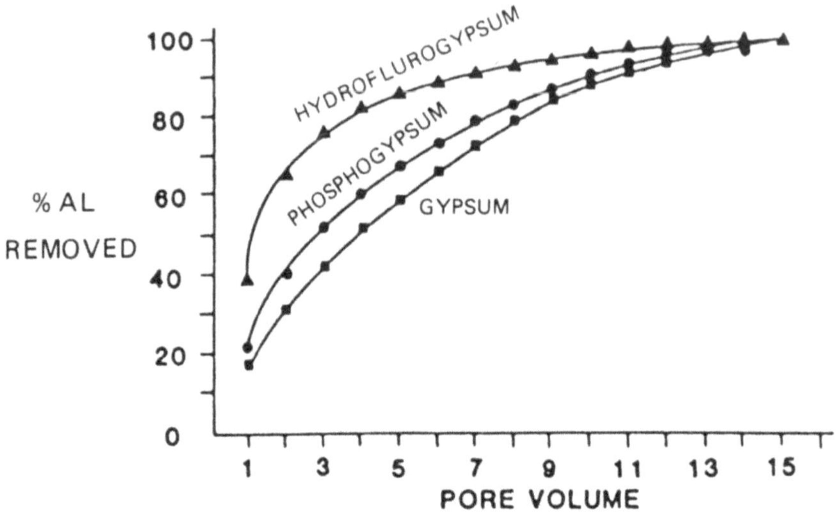

Figure 22. Percentage of Al leached from soil, amended with different gypsum products (Oates and Caldwell, 1985).

The adsorption/precipitation reactions involving SO_4^{2-} are useful in explaining the reduction in exchangeable Al observed with gypsum application, and so play an indirect but important role in plant growth. Since activities of Al and Ca species in solution relate best to plant growth, changes in exchangeable Al or Ca may or may not be sufficient to bring about a favorable effect as demonstrated by poor relationships between exchangeable Al and yield in many studies. As part of a complex set of reactions in a gypsum-treated soil, the adsorption/precipitation reactions can contribute to the reduction in solution Al^{3+} activity often observed. This is particularly important in the case of gypsum, since SO_4^{2-} is the anion accompanying Ca^{2+}. With other Ca salts, a similar change would not take place, and an increase in Al^{3+} activity would probably result from the increased ionic strength by exchange. A major advantage of gypsum application, then, is that it markedly increases ionic strength and solution Ca^{2+} activity while resulting in only a slight increase in solution Al^{3+} concentration and often a decrease in Al^{3+} activity.

Another proposed ameliorative mechanism is the leaching of Al from gypsum-treated soils. Using packed soil columns, Oates and Caldwell (1985) attributed growth response to leaching of Al after gypsum application (Figure 22). Gypsum was most effective when by-product materials such as hydrofluorogypsum and phosphogypsum were used because fluoride, which forms strong complexes with Al, is an impurity in each. However, other studies have not found leaching of Al to be substantial after application of gypsum (Pavan et al., 1984; Sumner et al. 1986). The leaching technique of Oates and Caldwell (1985; leaching rate of 3.6 cm h^{-1} for 30 h) would not generally be realistic for a field situation and so leaching of Al does not appear to be a primary mechanism in crop response to gypsum application.

A problem often encountered with gypsum application on highly weathered soils is the removal of K and Mg from the surface soil (Ritchey et al., 1980; Sumner et al., 1986). It would thus be advisable to include K and Mg amendments in any fertility program involving surface application of gypsum to highly weathered soils.

Although crop growth responses to gypsum have largely been attributed to chemical amelioration, physical amelioration may also play an important role. Radcliffe et al. (1986) concluded that improved aggregation and reduced subsoil strength resulted from greater root activity where gypsum application had produced a more favorable chemical environment. Work by Sumner et al. (1990) suggests that gypsum may result in a more direct physical improvement through improved flocculation and aggregation in subsoils where ionic strength and Ca^{2+} activity are initially very low.

V. Conclusions

Activities of Ca and Al species in solution probably give the best indication of a subsoil acidity problem. However, soil solution extraction (even saturated paste) is too time-consuming to be used as a routine test by soil-testing laboratories. Since activities of phytotoxic Al species are often difficult to determine accurately in complex solutions and since very low levels of Al can be toxic, problems remain while soil-solution Al is used as an indicator of subsoil acidity problems. Work with acid subsoils of warm, humid regions indicates that soil solution Ca^{2+} activity is consistently low in unamended subsoils where problems exist, suggesting that Ca^{2+} activity might be a good indicator of a problem and the potential for amelioration. A routine soil test relating to soil solution Ca^{2+} activity would greatly improve our ability to make appropriate recommendations to farmers and would be a valuable addition to other routine measurements such as Al and Ca saturation of the ECEC. Regression equations developed by Gillman and Sumner (1987) relating soil solution Ca^{2+} to exchangeable Ca^{2+} and solution electrical conductivity suggest that routine tests can be developed. Methodology such as that described by Cochrane and de Souza (1985) for measuring cation and anion exchange capacities in low-activity clay soils with solution strengths reflecting field conditions might also be useful.

Acid soils developed from acid-forming parent material rather than advanced pedogenic weathering must be managed differently because the specific nature of the problem is different. In these soils, Al toxicity is the major problem, not Ca insufficiency, and thus some form of extractable Al might be the best indicator. The surface charge characteristics of these subsoils should be examined in terms of how a particular amendment will alter the situation.

Deep liming is a possible method of amelioration in some situations, particularly if there is a hardpan that limits root penetration in addition to subsoil acidity. Of surface-applied amendments that have been studied in amelioration of subsoil acidity, gypsum has been shown to offer the greatest benefit in highly weathered soils. While the mechanisms of this response continue to be studied, the ability of gypsum to move quickly into the subsoil, increase solution Ca^{2+} activity, and often decrease Al^{3+} activity is important. With the soil surface charge characteristics being important in this response, gypsum has the most potential of ameliorating subsoil acidity where the soil pH is less than or relatively near pH_0. In addition, where gypsum is used, it is important to monitor topsoil levels of K and Mg, and applications of these nutrients will be necessary to offset leaching losses.

Access to a nearby source of by-product gypsum would substantially lower the cost of acid subsoil amelioration. Gypsum application rates up to 10 Mg ha^{-1} have been used with positive results in field studies, but lower rates might be sufficient and might even be necessary with salt-sensitive crops. With deep liming, surface application of gypsum or any other means of ameliorating acid subsoils, the long-term benefit will have to be weighed against initial investment costs.

References

Abruna-Rodriguez, F., J. Vicente-Chandler, E. Rivera, and J. Rodriguez. 1982. Effect of soil acidity factors on yields and foliar composition of tropical root crops. Soil Sci. Soc. Am. J. 46:1004–1007.

Adams, F., and P.J. Hathcock. 1984. Aluminum toxicity and calcium deficiency in acid subsoil horizons of two Coastal Plains soil series. Soil Sci. Soc. Am. J. 48:1305–1309.

Adams, F., and Z.F. Lund. 1966. Effect of chemical activity of soil solution aluminum on cotton root penetration on acid subsoils. Soil Sci. 101:193–198.

Adams, F., and B.L. Moore. 1983. Chemical factors affecting root growth in subsoil horizons of Coastal Plains soils. Soil Sci. Soc. Am. J. 47:99–102.

Adams, F., and Z. Rawajfih. 1977. Basaluminite and alunite: A possible cause of sulfate retention by acid soils. Soil Sci. Soc. Am. J. 41:686–692.

Alva, A.K., D.G. Edwards, C.J. Asher, and F.P.C. Blamey. 1986. Effects of phosphorus/aluminum molar ratio and calcium concentration on plant response to aluminum toxicity. Soil Sci. Soc. Am. J. 50:133–137.

Alva, A.K., M.E. Sumner, Y.C. Li, and W.P. Miller. 1989. Evaluation of three aluminum assay techniques for excluding aluminum complexed with fluoride or sulfate. Soil Sci. Soc. Am. J. 53:38–44.

Alva, A.K., M.E. Sumner, and A.D. Noble. 1988. Effects of fluoride on colorimetric measurement of solution aluminum. Soil Sci. Soc. Am. J. 52:374–378.

Anderson, D.L., and J.G. Hendrick. 1983. Subsoil lime injector. Soil Sci. Soc. Am. J. 47:337–339.

Awad, A.S., D.G. Edwards, and P.J. Milham. 1976. Effect of pH and phosphate on soluble soil aluminum and on growth and composition of kikuyu grass. Plant and Soil 45:531–542.

Bache, B.W., and W.M. Crooke. 1981. Interactions between aluminum, phosphorus, and pH in the response of barley to soil acidity. Plant and Soil 61:365–375.

Bache, B.W., and G.S. Sharp. 1976. Soluble polymeric hydroxy-aluminum ions in acid soils. J. Soil Sci. 27:167–174.

Barnhisel, R., and P.M. Bertsch. 1982. Aluminum. In: A.L. Page (ed.). Methods of Soil Analysis, Part 2. Agronomy 9:275–300. Am. Soc. Agron., Madison, WI.

Bartlett, R.J. 1982. Reactive aluminum in the Vermont soil test. Commun. Soil Sci. Plant Anal. 13:497–506.

Bartlett, R.J., and D.C. Riego. 1972. Toxicity of hydroxy-aluminum in relation to pH and phosphorus. Soil Sci. 114:194–200.

Bertsch, P.M., W.J. Layton, and R.I. Barnhisel. 1986. Speciation of hydroxy-aluminum solutions by wet chemical and aluminum-27 NMR methods. Soil Sci. Soc. Am. J. 50:1449–1454.

Black, A.S., and K.C. Cameron. 1984. Effect of leaching on soil properties and lucerne growth following lime and gypsum amendments to a soil with an acid subsoil. New Zealand J. Agr. Res. 27:195–200.

Blamey, F.P.C., D.G. Edwards, and C.J. Asher. 1983. Effects of aluminum, OH:Al and P:Al molar ratios, and ionic strength on soybean root elongation in solution culture. Soil Sci. 136:197–207.

Bloom, P.R., M.B. McBride, and R.M. Weaver. 1979a. Aluminum organic matter in acid soils: Buffering and solution aluminum activity. Soil Sci. Soc. Am. J. 43:488–493.

Bloom, P.R., M.B. McBride, and R.M. Weaver. 1979b. Aluminum organic matter in acid soils: Salt-extractable aluminum. Soil Sci. Soc. Am. J. 43:813–815.

Brenes, E., and R.W. Pearson. 1973. Root responses of three Gramineae species to soil acidity in an Oxisol and an Ultisol. Soil Sci. 116:295–302.

Brennan, W.C., and J.H. Axley. 1983. Influence of subsoil amelioration with lime, phosphorus, and potassium on the yield of soybeans and barley. Commun. Soil Sci. Plant Anal. 14:1119–1132.

Burstrom, Hans. 1952. Studies on growth and metabolism of roots. VIII. Calcium as a growth factor. Physiol. Plant 5:391–402.

Cameron, R.S., G.S.P. Ritchie, and A.D. Robson. 1986. Relative toxicities of inorganic aluminum complexes to barley. Soil Sci. Soc. Am. J. 50:1231–1236.

Cate, R.B., Jr., and L.A. Nelson. 1965. A rapid method for correlation of soil test analyses with plant response data. North Carolina Agric. Exp. Stn., International Soil Testing Series Tech. Bull. No. 1.

Chang, M.L., and G.W. Thomas. 1963. A suggested mechanism for sulfate adsorption by soils. Soil Sci. Soc. Am. Proc. 27:281–283.

Chao, T.T., M.E. Harward, and S.C. Fang. 1962. Soil constituents and properties in the adsorption of sulfate ions. Soil Sci. 94:276–283.

Chao, T.T., M.E. Harward, and S.C. Fang. 1963. Cationic effects of sulfate adsorption by soils. Soil Sci. Soc. Am. Proc. 27:35–38.

Clarkson, D.T., and J. Sanderson. 1971. Inhibition of the uptake and long-distance transport of calcium by aluminum and other polyvalent cations. J. Exp. Bot. 22:837–851.

Cochrane, T.T., and D.M.G. de Souza. 1985. Measuring surface charge characteristics in Oxisols and Ultisols. Soil Sci. 140:223–229.

Cochrane, T.T., J.G. Salinas, and P.A. Sanchez. 1980. An equation for liming acid mineral soils to compensate crop aluminum tolerance. Trop. Agric. (Trinidad) 57: 133–140.

Davenport, W.H. 1949. Determination of aluminum in the presence of iron. Anal. Chem. 21:710–711.

Emanuelsson, J. 1984. Root growth and calcium uptake in relation to calcium concentration. Plant and Soil 78:325–334.

Estrada, J., and G.A. Cummings. 1968. Effects of lime and phosphorus treatments in specific horizons of acid soil on growth and chemical content of corn. Agron. J. 60:447–450.

Evans, A., Jr., and L.W. Zelazny. 1987. Effects of sulfate additions on the status of exchangeable aluminum in a Cecil soil. Soil Sci. 143:410–417.

Evans, C.E., and E.J. Kamprath. 1970. Lime response as related to percent Al saturation, solution Al, and organic matter content. Soil Sci. Soc. Am. Proc. 34:893–896.

Farina, M.P.W., and P. Channon. 1988a. Acid-subsoil amelioration: I. A comparison of several mechanical procedures. Soil Sci. Soc. Am. J. 52:169–175.

Farina, M.P.W., and P. Channon. 1988b. Acid-subsoil amelioration: II. Gypsum effects on growth and subsoil chemical properties. Soil Sci. Soc. Am. J. 52:175–180.

Farina, M.P.W., M.E. Sumner, C.O. Plank, and W.S. Letzsch. 1980. Aluminum toxicity in corn at near neutral soil pH levels. J. Plant Nutrition. 2:683–697.

Fox, R.H. 1979. Soil pH, aluminum saturation, and corn grain yield. Soil Sci. 127:330–334.

Fox, R.L., and P.G.E. Searle. 1978. Phosphate absorption by soils of the tropics. In: M. Stelly (ed.). Diversity of Soils in the Tropics. Special Publication 34:97–119. Am. Soc. Agron., Madison, WI.

Foy, C.D. 1974. Effects of aluminum on plant growth. In: E.W. Carson (ed.). The Plant Root and Its Environment. pp. 601–642. Univ. Press of Virginia, Charlottesville.

Foy, C.D. 1984. Physiological effects of hydrogen, aluminum, and manganese toxicities in acid soil. In: F. Adams (ed.). Soil Acidity and Liming. Agronomy 12:57–97. Am. Soc. Agron., Madison, WI.

Gauch, H.G. 1972. Inorganic Plant Nutrition. Dowden, Hutchinson, and Ross, Inc., Stroudsburg, PA.

Gerard, C.J., H.C. Mehta, and E. Hinojosa. 1972. Root growth in a clay soil. Soil Sci. 114:37–49.

Gillman, G.P., and M.E. Sumner. 1987. Surface charge characterization and soil solution composition of four soils from the Southern Piedmont in Georgia. Soil Sci. Soc. Am. J. 51:589–594.

Gonzalez-Erico, E., E.J. Kamprath, G.C. Naderman, and W.V. Soares. 1979. Effect of depth of lime incorporation on the growth of corn on an Oxisol of central Brazil. Soil Sci. Soc. Am. J. 43:1155–1158.

Hammel, J.E., M.E. Sumner, and H. Shahandeh. 1985. Effect of physical and chemical profile modification on soybean and corn production. Soil Sci. Soc. Am. J. 49:1508–1511.

Hargrove, W.L. 1986. The solubility of aluminum-organic matter and its implication in plant uptake of aluminum. Soil Sci. 142:179–181.

Hargrove, W.L., and G.W. Thomas. 1981. Effect of organic matter on exchangeable aluminum and plant growth in acid soils. In: R.H. Dowdy et al. (eds.). Chemistry in the Soil Environment. pp. 151–166. ASA Spec. Publ. 40. ASA, Madison, WI.

Hargrove, W.L., and G.W. Thomas. 1984. Extraction of aluminum from aluminum-organic matter in relation to titratable acidity. Soil Sci. Soc. Am. J. 48:1458–1460.

Hodges, S.C. 1987. Aluminum speciation: A comparison of five methods. Soil Sci. Soc. Am. J. 51:57–64.

Howard, D.D., and F. Adams. 1965. Calcium requirement for penetration of subsoils by primary cotton roots. Soil Sci. Soc. Am. Proc. 29:558–562.

Hsu, P.H. 1963. Effect of initial pH, phosphate, and silicate on the determination of aluminum with aluminon. Soil Sci. 96:230–238.

Hue, N.V., F. Adams, and C.E. Evans. 1985. Sulfate retention by an acid BE horizon of an Ultisol. Soil Sci. Soc. Am. J. 49:1196–1200.

Hue, N.V., G.R. Craddock, and F. Adams. 1986. Effect of organic acids on aluminum toxicity in subsoils. Soil Sci. Soc. Am. J. 50:28–34.

James, B.R., C.J. Clark, and S.J. Riha. 1983. An 8-hydroxyquinoline method for labile and total aluminum. Soil Sci. Soc. Am. J. 47:893–897.

Jardine, P.M., and L.W. Zelazny. 1987a. Influence of organic anions on the speciation of mononuclear and polynuclear aluminum by ferron. Soil Sci. Soc. Am. J. 51:885–889.

Jardine, P.M., and L.W. Zelazny. 1987b. Influence of inorganic anions on the speciation of mononuclear and polynuclear aluminum by ferron. Soil Sci. Soc. Am. J. 51:889–892.

Jenny, H. 1961. Reflections on the soil acidity merry-go-round. Soil Sci. Soc. Am. Proc. 25:428–432.

Jones, R.W. Wyn, and O.R. Lunt. 1967. The function of calcium in plants. Bot. Rev. 33:407–425.

Kamprath, E.J. 1970. Exchangeable aluminum as a criterion for liming leached mineral soils. Soil Sci. Soc. Am. Proc. 34:252–254.

Khalid, R.A., and J.A. Silva. 1980. Residual effect of calcium silicate on pH, phosphorus, and aluminum in a tropical soil profile. Soil Sci. Plant Nutr. 26:87–98.

Khan, S.U. 1969. Interaction between the humic acid fraction of soils and certain metallic cations. Soil Sci. Soc. Am. Proc. 33:851–854.

Kinraide, T.B., and D.R. Parker. 1987. Non-phytotoxicity of the aluminum sulfate ion, $AlSO_4^+$. Physiol. Planta. 71:207–212.

Kirkby, E.A. 1979. Maximizing calcium uptake by plants. Commun. Soil Sci. Plant Anal. 10:89–113.

Lance, J.C., and R.W. Pearson. 1969. Effect of low concentrations of aluminum on growth and water and nutrient uptake by cotton roots. Soil Sci. Soc. Am. Proc. 33:95–98.

Lee, J., and M.W. Pritchard. 1984. Aluminum toxicity expression on nutrient uptake, growth, and root morphology of *Trifolium repens* L. cv. 'Grasslands Huia.' Plant and Soil 82:101–116.

Lemus-Grob, F. 1985. Effects and behavior of gypsum in an Ultisol. M.S. thesis. University of Georgia, Athens.

Lindsay, W.L. 1979. Chemical Equilibria in Soils. Wiley, New York.

Lund, Z.F. 1970. The effect of calcium and its relation to several cations in soybean root growth. Soil Sci. Soc. Am. Proc. 34:456–459.

McCray, J.M., M.E. Sumner, D.E. Radcliffe, and R.L. Clark. 1989. Effects of subsoiling, lime, and gypsum on soil strength, soil acidity, cation movement, and cotton growth. Manuscript submitted for publication.

McKenzie, R.C., and M. Nyborg. 1984. Influence of subsoil acidity on root development and crop growth in soils of Alberta and northeastern British Columbia. Can. J. Soil Sci. 64:681–697.

Moore, C.S., and G.S.P. Ritchie. 1988. Aluminum speciation and pH of an acid soil in the presence of fluoride. J. Soil Sci. 39:1–8.

Moschler, W.W., G.D. Jones, and G.W. Thomas. 1960. Lime and soil acidity effects on alfalfa growth in a red-yellow podzolic soil. Soil Sci. Soc. Am. Proc. 24:507–509.

Noble, A.D., M.V. Fey, and M.E. Sumner. 1988a. Calcium-aluminum balance and the growth of soybean roots in nutrient solutions. Soil Sci. Soc. Am. J. 52:1651–1656.

Noble, A.D., and M.E. Sumner. 1988. Ca and Al interactions and soybean growth in nutrient solutions. Commun. Soil Sci. Plant Anal. 19:1119–1131.

Noble, A.D., M.E. Sumner, and A.K. Alva. 1988b. Comparison of aluminon and 8-hydroxyquinoline methods in the presence of fluoride for assaying phytotoxic aluminum. Soil Sci. Soc. Am. J. 52:1059–1063.

Noble, A.D., M.E. Sumner, and A.K. Alva. 1988c. The pH dependency of aluminum phytotoxicity alleviation by calcium sulfate. Soil Sci. Soc. Am. J. 52:1398–1402.

Nordstrom, D.K. 1982. The effect of sulfate on aluminum concentrations in natural waters: Some stability relations in the system Al_2O_3–SO_3–H_2O at 298 K. Geochim. Cosmochim. Acta 46:681–692.

Oates, K.M., and A.G. Caldwell. 1985. Use of by-product gypsum to alleviate soil acidity. Soil Sci. Soc. Am. J. 49:915–918.

Oates, K.M., and E.J. Kamprath. 1983a. Soil acidity and liming: I. Effect of the extracting solution cation and pH on the removal of aluminum from acid soils. Soil Sci. Soc. Am. J. 47:686–689.

Oates, K.M., and E.J. Kamprath. 1983b. Soil acidity and liming: II. Evaluation of using aluminum extracted by various chloride salts for determining lime requirements. Soil Sci. Soc. Am. J. 47:690–692.

Okura, T., K. Goto, and T. Yotuyanagi. 1962. Forms of aluminum determined by an 8-quinolinolate extraction method. Anal. Chem. 34:581–582.

Parfitt, R.L. 1978. Anion adsorption by soils and soil materials. Advan. Agron. 30:1–41.

Parker, D.R., T.B. Kinraide, and L.W. Zelazny. 1988. Aluminum speciation and phytotoxicity in dilute hydroxy-aluminum solutions. Soil Sci. Soc. Am. J. 52:438–444.

Pavan, M.A., and F.T. Bingham. 1982. Toxicity of aluminum to coffee seedlings grown in nutrient solution. Soil Sci. Soc. Am. J. 46:993–997.

Pavan, M.A., F.T. Bingham, and P.F. Pratt. 1982. Toxicity of aluminum to coffee in Ultisols and Oxisols amended with $CaCO_3$, $MgCO_3$, and $CaSO_4 \cdot 2H_2O$. Soil Sci. Soc. Am. J. 46:1201–1207.

Pavan, M.A., F.T. Bingham, and P.F. Pratt. 1984. Redistribution of exchangeable calcium, magnesium, and aluminum following lime or gypsum applications to a Brazilian Oxisol. Soil Sci. Soc. Am. J. 48:33–38.

Radcliffe, D.E., R.L. Clark, and M.E. Sumner. 1986. Effect of gypsum and deep-rooting perennials on subsoil mechanical impedance. Soil Sci. Soc. Am. J. 50:1566–1570.

Ragland, J.L., and N.T. Coleman. 1959. The effect of soil solution aluminum and calcium on root growth. Soil Sci. Soc. Am. Proc. 23:355–357.

Rajan, S.S.S. 1978. Sulfate adsorbed on hydrousalumina, ligands displaced, and changes in surface charge. Soil Sci. Soc. Am. J. 42:39–44.

Reeve, N.G., and M.E. Sumner. 1970a. Effects of aluminum toxicity and phosphorus fixation on crop growth on Oxisols in Natal. Soil Sci. Soc. Am. Proc. 34:263–267.

Reeve, N.G., and M.E. Sumner. 1970b. Lime requirements of Natal Oxisols based on exchangeable aluminum. Soil Sci. Soc. Am. Proc. 34:595–598.

Reeve, N.G., and M.E. Sumner. 1972. Amelioration of subsoil acidity in Natal Oxisols by leaching of surface-applied amendments. Agrochemophysica 4:1–5.

Ritchey, K.D., D.M.G. Souza, E. Lobato, and O. Correa. 1980. Calcium leaching to increase rooting depth in a Brazilian savannah Oxisol. Agron. J. 72:40–44.

Saigusa, M., S. Shoji, and T. Takahashi. 1980. Plant root growth in acid andosols from northeastern Japan: 2: Exchange acidity Y_1 as a realistic measure of aluminum toxicity potential. Soil Sci. 130:242–250.

Sartain, J.B., and E.J. Kamprath. 1977. Effect of soil Al saturation on nutrient concentration of soybean tops, roots, and nodules. Agron. J. 69:843–845.

Sheppard, L.J., and M.J.S. Floate. 1984. The effects of soluble-Al on root growth and radicle elongation. Plant and Soil 80:301–306.

Simpson, J.R., A. Pinkerton, and J. Lazdovskis. 1979. Interacting effects of subsoil activity and water on the root behavior and shoot growth of some genotypes of lucerne (*Medicago sativa* L.). Aust. J. Agric. Res. 30:609–619.

Singh, S.S. 1982. Effect of chloride and sulfate anions on the chemical characteristics of some acid soils. Can. J. Soil Sci. 62:549–557.

Sposito, G. 1985. Chemical models of weathering in soils. In: J.I. Drever (ed.). The Chemistry of Weathering. D. Reidel Publishing Company, New York.

Sposito, G., and S.V. Mattigod. 1980. GEOCHEM: A computer program for the calculation of chemical equilibria in soil solutions and other natural water systems. Kearney Foundation of Soil Science, University of California, Riverside.

Stillwell, T.C., and T.G. Arscott. 1978. The relationship between fluoride-titratable (reactive) soil aluminum and plant growth. Soil Sci. 125:28–33.

Sumner, M.E. 1970. Aluminum toxicity-growth limiting factor in some Natal sands. Proc. S. A. Sug. Tech. Assoc. 44:1–6.

Sumner, M.E., D.E. Radcliffe, M. McCray, E. Carter, and R.L. Clark. 1990. Gypsum as an ameliorant for subsoil hardpans. Soil Technol. In press.

Sumner, M.E., H. Shahendeh, J. Bouton, and J. Hammel. 1986. Amelioration of an acid soil profile through deep liming and surface application of gypsum. Soil Sci. Soc. Am. J. 50:1254–1258.

Tanaka, A., T. Tadano, K. Yamamoto, and N. Kanamura. 1987. Comparison of toxicity to plants among Al^{3+}, $AlSO_4^+$, and Al-F complex ions. Soil Sci. Plant Nutr. 33:43–55.

Tanaka, W., and F.W. Woods. 1972. Root and root hair growth in relation to supply and internal mobility of calcium. Bot. Gaz. 133:29–34.

Taylor, H.M., G.M. Roberson, and J.J. Parker, Jr. 1966. Soil strength-root penetration relations for medium to coarse textured soil materials. Soil Sci. 102:18–22.

Thomas, G.W., and W.L. Hargrove. 1984. The chemistry of soil acidity. In: F. Adams (ed.). Soil Acidity and Liming. Agronomy 12:57–97. Am. Soc. Agron., Madison, WI.

Uehara, G., and G.P. Gillman, 1981. The mineralogy, chemistry, and physics of tropical soils with variable charge clays. Westview Press, Boulder, CO.

White, R.E., L.O. Tiffin, and A.W. Taylor. 1976. The existence of polymeric complexes in dilute solutions of aluminum and orthophosphate. Plant and Soil 45:521–529.

Wright, J.R., and M. Schnitzer. 1963. Metallo-organic interactions associated with podzolization. Soil Sci. Soc. Am. Proc. 27:171–176.

Wright, R.J., V.C. Baligar, and S.F. Wright. 1987. Estimation of phytotoxic aluminum in soil solution using three spectrophotometric methods. Soil Sci. 144:224–232.

Wright, R.J., J.L. Hern, V.C. Baligar, and O.L. Bennett. 1985. The effect of surface applied amendments on barley root growth in an acid subsoil. Commun. Soil Sci. Plant Anal. 16:179–192.

Wright, R.J., and S.F. Wright. 1987. Effects of aluminum and calcium on the growth of subterranean clover in Appalachian soils. Soil Sci. 143:341–348.

Zelazny, L.W., and P.M. Jardine. 1988. Surface reactions of aqueous aluminum. In: G. Sposito (ed.). Environmental Chemistry of Aluminum. CRC Press, Boca Raton, FL.

Application of Electron Spin Resonance (ESR) Spectroscopy in Soil Chemistry

N. Senesi

Advances in Soil Science, Volume 14

I. Introduction

Organic free radicals and most transition metal ions are characterized by the presence in their structure of one or more unpaired electron(s), that is, they are paramagnetic, and may produce electron spin resonance (ESR) signals in both the free state and any organic and inorganic association in which they maintain unpaired electron(s), in either solid or solution state.

The paramagnetic properties of these species have stimulated numerous studies, since organic free radicals are indigenous to humic substances from soil and related materials, and paramagnetic metal ions are widespread in the soil system in a variety of soluble and insoluble organic-and mineral-associated forms.

Because humic substances (HS) represent the most abundant and most chemically and biologically active fraction of soil organic matter and metal ions are involved in practically all phenomena that occur in the soil, particularly at the solution–solid interfaces, information obtained by organic and metal ESR probes is of great importance in many soil chemical investigations.

The intention of this review is to first examine the basic physical principles, applicable theory and methodology, and some intrinsic and experimental limitations of ESR spectroscopy. After that, a review and evaluation are furnished of significant results and the impact of ESR applications to the study of soil HS and similar materials, of their natural and laboratory-modeled interactions with organic chemicals and metal ions, and of metal associations with soil and synthetic inorganic oxides and hydrous oxides and clay minerals in the presence or absence of organics. The review ends with suggestions and recommendations for future extension and novel applications of the ESR spectroscopy and germane magnetic techniques ENDOR and spin-echo.

II. Basic Principles and Methodology of ESR Spectroscopy

A. Physical Principles

The basic phenomenon underlying ESR spectroscopy is the Zeeman effect, which involves the interaction between the spin of an unpaired electron and an external magnetic field. As in other types of spectroscopy, the ESR signal results from the absorption of energy from an incident electromagnetic "wave," which

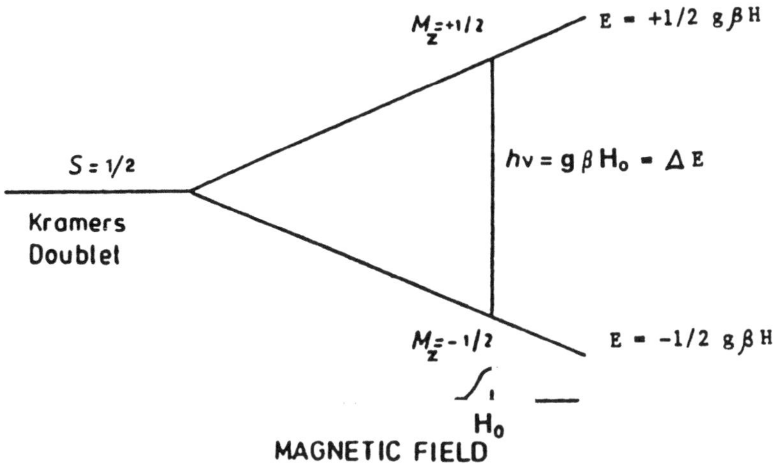

Figure 1. Effect of an applied magnetic field of strength H on the energy levels (E) of the two spin states of an electron. ESR transition for a Kramers doublet at $\nu = g\beta H_0/h$.

induces a transition of a molecule (or portion of thereof) from a state of lower energy to one of higher energy.

If a static magnetic field of strength H, in Gauss (G), is applied parallel to the z-axis, the Zeeman interaction leads to an energy for the electronic magnetic dipole given by

$$E = -\mu_z H \tag{1}$$

where μ_z, the electronic magnetic moment with respect to the conventional reference z-axis, is given by

$$\mu_z = -g\beta M_z \tag{2}$$

Combination of equations 1 and 2 gives, for the energy of the unpaired electron,

$$E = -g\beta H\, M_z = \pm 1/2\; g\beta\; H \tag{3}$$

where M_z is the component of the electron-spin angular momentum in the direction of the applied magnetic field, H (z-axis), β is the Bohr magneton, and g is the magnetogyric ratio or spectroscopic splitting factor (g-factor or g-value), which takes into account the orbital contribution to the spin-magnetic moment.

Because M_z may have values of $+1/2$ and $-1/2$, the alignment of the spin-magnetic moment may be either with the magnetic field direction (high energy) or against it (low energy). Thus, the spin states are not energetically equivalent in an applied magnetic field H, where the energy difference between the two spin states is

$$\Delta E = g\beta H \tag{4}$$

and increases linearly with the intensity of H (Fig. 1).

Magnetic dipole transitions between the two spin states, from lower level to upper level, can therefore be readily induced by incident electromagnetic radiation, that is, by applying an alternating magnetic field of frequency ν perpendicularly to the static magnetic field H_0, provided that

$$h\nu = \Delta E = g\beta H_0 \tag{5}$$

This is known as the "resonance condition." The ESR transition for a two-level system with $S = 1/2$ (Kramers doublet) is shown in Figure 1.

A net absorption of energy occurs from the alternating field, however, only when the population of the lower level is greater than that of the upper level. The population difference in an applied static field in thermal equilibrium and in the absence of an alternating field is given by the ratio

$$\frac{N_+}{N_-} = e^{\frac{-g\beta H_0}{kT}} \tag{6}$$

where N_+ and N_- refer to the population of the $M = +1/2$ and $M = -1/2$ levels, respectively, T is the temperature, and k is the Boltzmann constant. Provided the magnitude of the alternating field is not too large, this population difference is maintained because electrons in the $M = +1/2$ state lose energy to their surroundings (spin-lattice relaxation phenomenon) and return to the $M = -1/2$ state. Thus, the energy levels are not sharply defined, so that the sample absorbs energy over a range of field values, giving an absorption line shape consisting of a curve of some finite width, rather than an infinitely narrow peak at some resonating field strength, H_0.

The measurement of this absorption of energy is the basis of ESR spectroscopy.

B. Instrumentation

ESR spectrometers operate in the microwave region of the electromagnetic spectrum (ν = 9–35 GHz), the X-band spectrometers (typical microwave frequency around 9.5 GHz) being the most common (Fig. 2). The sample is placed in a chamber called a "microwave cavity" and irradiated with microwaves whose frequency is held constant to within 1 part in 10^6 by a vacuum tube oscillator called klystron. The klystron, that is, the source of the alternating radiation field, is coupled by a waveguide to the resonant cavity in which the sample is placed and locked electronically to the resonant frequency. A circulator, incorporated between the klystron and the cavity, ensures optimum transfer of the microwave power from the klystron to the cavity and from the cavity to the detector. The cavity is coupled to the waveguide by a small iris, whose effective diameter can be controlled by the movement of a suitable device. When the iris diameter is such that all the incident microwaves are absorbed by the cavity, the cavity is said to be "critically coupled." The spectrometer is usually operated as closely as possible to this condition.

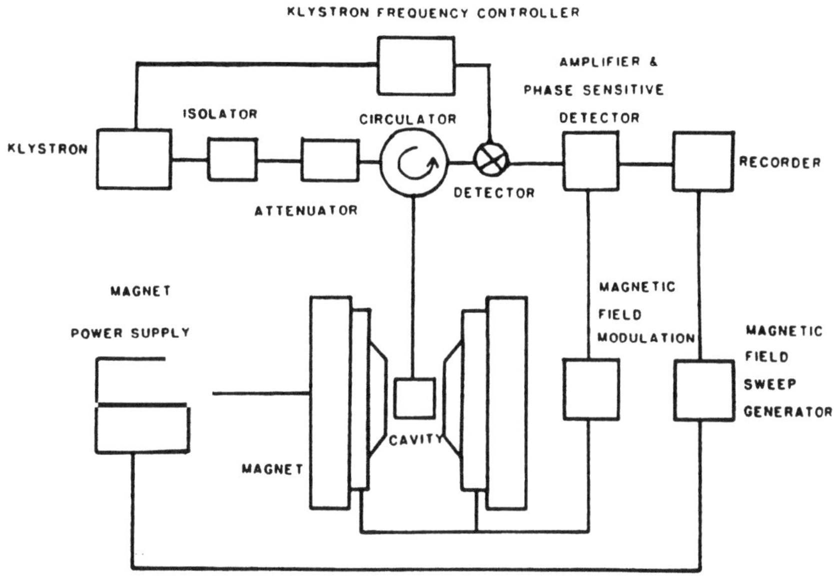

Figure 2. The basic ESR spectrometer.

The static, or DC magnetic field (practical upper limit 10 to 20 kG for commercial spectrometers) is generated by large electromagnets flanking the microwave cavity. The applied field is swept linearly through the resonance conditions (Eq. 5) at a rate that depends on the characteristics of both the sample and the instrumentation. A typical sweep rate may be around 200 G/min. Most ESR spectrometers provide for repetitive sweeping of the magnetic field through the resonance, allowing the employment of a computer to average transients. Sensitivity is thus increased, since the signals add proportionally to the number of sweeps, while the noise adds as the square root.

When resonance occurs, the absorption of microwave energy by the sample changes the effective coupling of the cavity to the waveguide and hence the amount of microwave power reflected from the cavity. Therefore, a change occurs in the microwave-power level at the detector. The most common method of detecting this change is to modulate the DC by slowly varying with time the external magnetic field at 100 kHz and to employ phase-sensitive detection. The output of the detector–amplifier system thus appears as the first derivative of the absorption and is usually recorded on a chart-recorder as a function of the external magnetic field, H (Fig. 3).

In actual practice, observation of resonance requires net absorption of energy, which can occur only in molecules with net spin angular momentum, including transition elements, such as copper, iron, manganese, vanadium, and molybdenum, and free radicals, that is, molecules containing a single, unpaired electron.

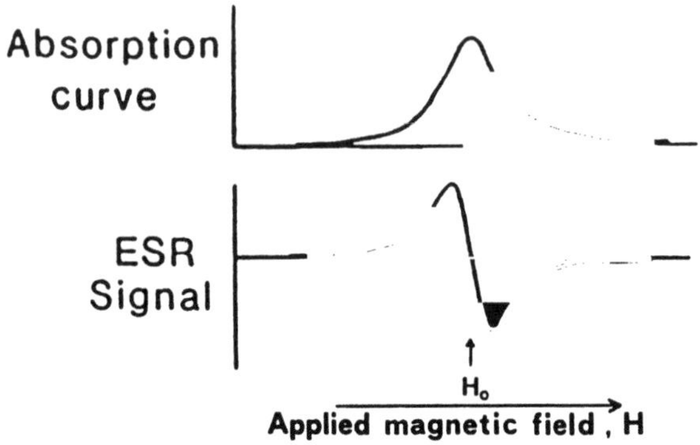

Figure 3. Absorption of microwave energy as a function of applied, external magnetic field of strength H. Actual absorption curve (above) and ESR signal recorded from the spectrometer, that is, the first derivative of the absorption curve (below).

A detailed presentation of experimental techniques of ESR spectroscopy and of the factors affecting ESR spectral sensitivity and resolution may be found in Poole (1967), Alger (1968), and Ingram (1969).

C. Methodology

If the ESR signal consisted of a single line similar to that in Figure 3, the technique would be of limited use. Fortunately, the ESR signal is extremely sensitive to the conditions of the local environment about the absorbing electrons. The effective magnetic field experienced by an electron (H_{eff}) is the result of two terms: the applied magnetic field generated by the spectrometer magnet (H_{appl}) and the local perturbations of the field produced by the electron's environment (H_{loc}), that is,

$$H_{eff} = H_{appl} + H_{loc} \tag{7}$$

In definitive, H_{eff} determines the separation of the Zeeman energy levels and the resonant field strength H_0, thus, depends on environmental conditions in the vicinity of the electron. This, in turn, affects both the position of the ESR signal and the shape of the spectrum, thus providing considerable information about the molecular structure of the system studied.

1. The Spin Hamiltonian

The simple observation of an ESR spectrum alone does not directly furnish much chemical and structural information on the sample studied. To obtain such information, once the ESR spectrum is measured, it is necessary to represent it mathematically and find values of parameters that can be related to chemical and structural details of the sample, such as free radical nature, metal species and oxidation state, site symmetry and bonding, and other factors.

The quantum mechanical operator "spin Hamiltonian" is a convenient means of representing an ESR spectrum and is simply the sum of a number of terms involving these parameters and the electron and nuclear spin operators. If a spin Hamiltonian can be fitted to the experimental spectrum, we can use the parameters so obtained to make deductions about the chemical and structural properties of the sample.

A brief development and discussion of the spin Hamiltonian for mononuclear copper (II) ions is furnished here, as a relevant example to soil chemistry application of ESR spectroscopy. More detailed treatments can be found in Wertz and Bolton (1972), Swartz et al. (1972), Carrington and McLachlan (1967), and Abragam and Bleaney (1970).

The spin Hamiltonian for mononuclear copper (II) ions can be written as

$\mathrm{H} = (g_x B_x \bar{S}_x + g_y B_y \bar{S}_y + g_z B_z \bar{S}_z)$	the electron Zeeman interaction
$+ (A_x \hat{S}_x \hat{I}_x + A_y \hat{S}_y \hat{I}_y + A_z \hat{S}_z \hat{I}_z)$	the hyperfine interaction
$+ \Sigma_i (A_{ix} \hat{S}_x \hat{I}_{ix} + A_{iy} \hat{S}_y \hat{I}_{iy} + A_{iz} S_z I_{sz})$	the ligand superhyperfine interaction
$+ P_{\parallel}\{[\hat{I}_{\mathrm{z}}^2 - 1/3\, I\, (I + 1)] + 1/3\, (\hat{I}_{\mathrm{x}}^2 - \hat{I}_{\mathrm{y}}^2)\}$	the quadrupole interaction
$+ \beta\, (g_x^I B_x \hat{I}_x + g_y^I B_y \hat{I}_y + g_z^I B_z \hat{I}_z)$	the nuclear Zeeman interaction
(terms of the spin Hamiltonian containing the various parameters and operators)	(type of interaction corresponding to each term of the spin Hamiltonian)

This expression assumes that all tensors and matrix axes are coincident. Being a quantum mechanical operator, the spin Hamiltonian includes terms containing the electron spin operators $\hat{S}_x$, $\hat{S}_y$, and $\hat{S}_z$; the nuclear spin operators $\hat{I}_x$, $\hat{I}_y$, and $\hat{I}_z$, and the nuclear spin operators of the ligands $\hat{I}_{ix}$, $\hat{I}_{iy}$, and $\hat{I}_{iz}$. It also involves the components of the flux density, B, of the applied magnetic field B_x, B_y, and B_z. The magnetic properties of the ion are determined by the parameters of the spin Hamiltonian, namely, the components of the g, $\underline{\mathrm{A}}$, g^I, and $\underline{\mathrm{A}}_i$ matrices and $\underline{\mathrm{P}}$ tensor.

The most important term in any spin Hamiltonian is the "electron Zeeman" interaction that arises from unpaired electrons and external magnetic field. It takes into account the g-value at which resonance occurs for a particular orientation of the magnetic field with respect to the principal axes, x, y, z, of the g matrix. Generally, Cu (II) and often VO(II) ions exhibit axial or nearly axial symmetry with principal g-values, $g_{\parallel}$ (the g-value along the z or symmetry axis) and $g_{\perp}$ (the g-value in the x-y plane).

The "nuclear hyperfine" interaction arises from the electron and its nucleus. The nuclear magnetic moment, therefore, alters H_{eff}, changing H_0. Only nuclei

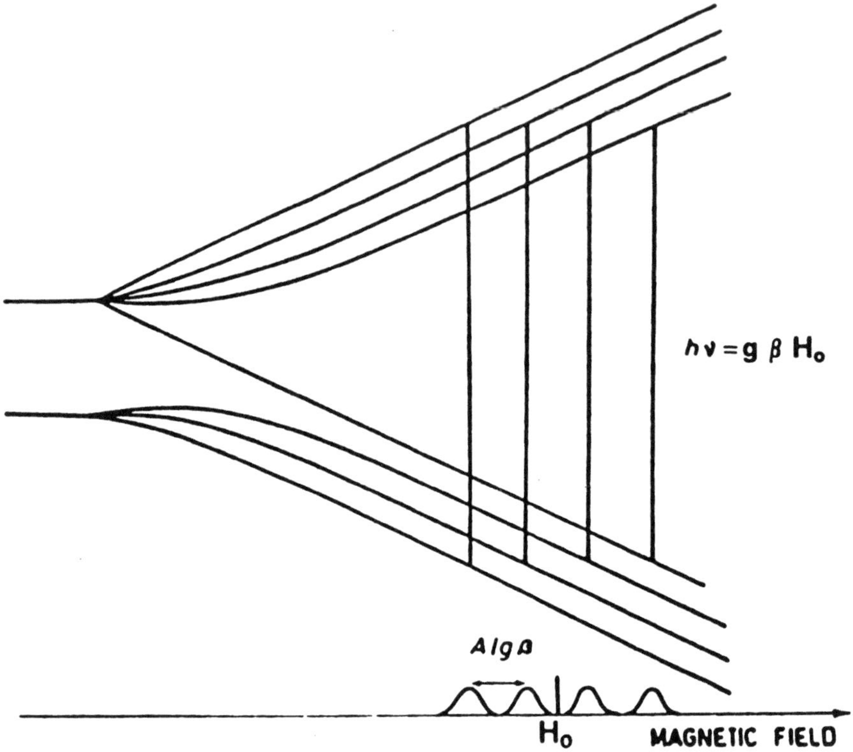

Figure 4. The ESR transition expected from an ion with $S = 1/2$ and $I = 3/2$ [e.g., Cu(II) ion]. The hyperfine splitting is $A/g\beta$.

with nonzero spin ($I \neq 0$), for example, Cu (I = 3/2), Mn (I = 5/2), V (I = 7/2), exhibit this type of interaction. Each electron spin level is split into $2I+1$ components, for example, four components for Cu, six for Mn, and eight for V. Thus, the electron resonance line is split into the same number of approximately equally spaced components, as shown in Figure 4 for Cu (II) ion. The splitting of the hyperfine components is, in general, $A/g\beta$ as a first approximation, where A is the magnitude of the nuclear hyperfine interaction, that is, the "hyperfine coupling constant." As g, A also has a directional dependence and $\underline{A}$ is another interaction matrix that can be diagonalized by an appropriate choice of axes.

The "ligand superhyperfine" interaction arises from the magnetic moments of ligand nuclei and the electron spin. This interaction can also be expressed in terms of a matrix, but it may have different principal axes and different components for each ligand nucleus, thus leading to very complex spectra, particularly if the ligands are not identical. The most common nucleus giving rise to superhyperfine structure in the ESR spectra of Cu(II) complexes is ^{14}N, which has a nuclear spin, $I = 1$. Thus, each hyperfine line is split into three approximately equally spaced components of equal intensity.

The "nuclear quadrupole" interaction arises from the nuclear quadrupole moment (since, for the Cu nucleus, $I = 3/2$) and the electric field gradient. The interaction shifts the relative positions of the nuclear hyperfine levels and thus alters the observed hyperfine splittings. It also "mixes" the hyperfine levels, so that "forbidden" transitions can take place. Generally, quadrupole effects are very small and less readily observed in powder and frozen solution spectra.

The "nuclear Zeeman" interaction arises from the magnetic moment of the copper nucleus and the external magnetic field. The effect of this term is expected to be quite small (less than 1/1000 that of the term in g; Abragam and Bleaney, 1970).

2. Spin Hamiltonian Parameters

To determine accurately the spin Hamiltonian parameters, the experimental spectrum is compared to a computer-simulated spectrum, which is calculated with the use of trial parameters and the expressions for the magnetic fields at which transitions occur. A new set of trial parameters is chosen, the spectrum is simulated again, and its is then compared to the experimental one. This process is repeated until satisfactory agreement is found between the two spectra.

In practice, the elaborate computer simulation procedure can be avoided by the relatively simple, direct computation of g-values and hyperfine constants, A, from the experimental spectrum, according to the standard equations (Wertz and Bolton, 1972)

$$g = \frac{h\nu}{\beta H_0} = 0.714484 \frac{\nu}{H_0} \tag{9}$$

where ν (MHz) is the accurately measured value of the microwave frequency used, H_0 is the value of the field at which the resonance is centered, h is the Planck's constant, and β is the Bohr magneton;

$$A\,(\text{MHz}) = 2.80247 \left(\frac{g}{g_e}\right) a \tag{10}$$

where $g_e = 2.00232$ (g-value for the spin of the free electron), g is calculated by equation 9, and a, in Gauss (1 Gauss = 10^{-4} mTesla, mT), is the hyperfine constant computed on the experimental spectrum; and

$$A\,(\text{cm}^{-1}) = \frac{A}{c} = 0.333564\ 10^{-4}\, A(\text{MHz}) \tag{11}$$

where c is the speed of light in vacuum.

3. Interpretation of Spectral Parameters

Once the ESR parameters are obtained, they do not give any direct insight into the structural and bonding features of the responsible paramagnetic species. To obtain this, it is necessary to relate g and A values to the electronic structure and symmetry of the site of the species studied by using a simple "crystal field" model approximation or more elaborate "ligand field" and "molecular orbital" calculations.

To do so, one must derive analytic expressions for the g-values and hyperfine constants in terms of the splittings between the energy levels and the admixtures of the appropriate wave functions, a process achieved by applying a reduced form of the total Hamiltonian of the system to wave functions. An extensive discussion on the methods of derivation of the reduced Hamiltonian and the calculations thereof is beyond the scope of this chapter and can be found in Abragam and Bleaney (1970), among others.

The considerably complex task of performing elaborated mathematical calculations based on a strict physicochemical approach may be completely or partially avoided in some instances. Empirical comparison and correlation of ESR spectral parameters determined by the accurate evaluation of experimental spectra with ESR data referred to in the scientific literature on similar model or synthetic molecular systems or natural compounds (e.g., metal proteins, coals, organic polymers, etc.) often furnishes the information of chemical and structural interest to soil scientist.

D. Limitations of ESR Spectroscopy

1. Line Resolution

The major limitation of an ESR experiment is the resolution, which is determined by the width of the component lines. These component lines may overlap to such an extent that information is lost, and this situation is most likely to occur when powders or frozen solutions are studied. The width of a resonance line is determined by two mechanisms, namely, homogeneous and inhomogeneous broadening.

Generally, the most important homogeneous broadening mechanism is "spin-lattice relaxation," that is, the mechanism by which an electron that has absorbed energy, and therefore is occupying a higher energy level, loses energy to its environment (i.e., the "lattice"), in order to return to the ground state. This process, although spontaneous, requires a time, called "spin-lattice relaxation time," that depends on the rate at which the electron may lose energy, which tends to be strongly temperature dependent. Mechanisms such as spin-lattice relaxation therefore determine the width of the resonance line by affecting the lifetime of the excited state of the system. Resonance is therefore observed over a range of H_{appl}, and the ESR signal has a finite width. Larger values of spin-lattice relaxation time result in more narrowly defined Zeeman energy levels, and transitions therefore occur over a narrower range of H_{appl}, resulting in narrower line widths.

"Microwave-power saturation" effects are a consequence of a spin system that has a finite relaxation time. The observation of an ESR line depends on the existence of a population difference between the ground and excited electron spin levels, as we considered previously in this review, and the intensity of the resonance absorption depends on the magnitude of the population difference. If a sufficiently intense microwave frequency field, that is, a high microwave power level, is applied, the number of electrons excited from the ground state may equal the number that return, so that the populations of the two levels tend to equalize and the absorption of microwave energy decreases. This phenomenon results in a decreased signal intensity and an increased line width, as the microwave power increases. In the absence of saturation, the signal intensity increases as the square root of the microwave power.

Inhomogeneous broadening arises from nonuniformities in the magnetic field throughout the sample, as a result of other neighboring paramagnetic species or the magnetic moments of neighboring nuclei, that is, unresolved fine, hyperfine, or superhyperfine structure, or from dipolar interactions between unlike spins. The result is a spectral distribution of individual resonant lines or spin packets, which merge into a single overall line or envelope. This interaction is random in direction and is often referred to as the "spin–spin interaction." Two particular cases can be mentioned (Abragam and Bleaney, 1970). First, if the paramagnetic species are identical, the effect is that of an energy exchange between the species, which broadens and shortens the lifetime of the individual species in a given state. Second, the effect of nearby nuclei with magnetic moments may be to inhomogeneously broaden the resonance lines, so that the hyperfine interaction is not resolved. The latter case is often observed for organic free radicals in humic substances, while the former may represent a major factor in the nonobservation of spectra for highly concentrated copper systems tumbling in solution at room temperature (RT). A third form of inhomogeneous broadening (*g*-strain broadening) arises from local variations in the position of neighboring species, which gives rise to variations in the *g*-values of the individual paramagnetic species and thus a variation in the magnetic field at which the resonance occurs. Two magnetic resonance techniques, namely, electron nuclear double resonance (ENDOR) and spin-echo spectrometry, can be used to overcome the difficulties caused by inhomogeneous broadening.

2. Temperature Effects

The temperature at which ESR measurements are made affects not only the sensitivity of the measurement, but also the type of information obtained. A major disadvantage of working at RT in aqueous systems is that water has an extremely high dielectric loss at microwave frequencies. Thus, small samples must be used to maintain sensitivity, and the sample must be placed in a capillary tube or a specially constructed flat solution cell, to minimize dielectric losses sufficiently. For these reasons, most ESR measurements are made at reduced temperature on frozen solutions, with either liquid nitrogen (bp 77 K) or liquid helium (bp 4.2 K).

Large metal–organic molecules (e.g., metal–humic substance complexes), however, cannot tumble rapidly in solution, thus the ESR spectrum at RT is similar to that observed for powders or frozen solutions, with the previously mentioned disadvantages of lower sensitivity.

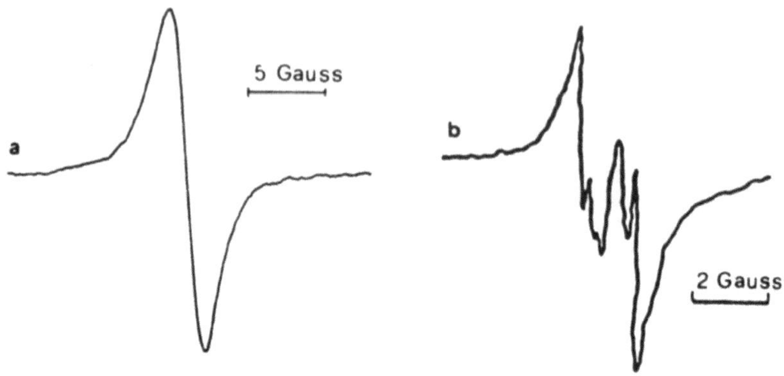

Figure 5. ESR spectra of humic substances. (a, Reproduced from N. Senesi and C. Steelink, "Application of ESR spectroscopy to the study of humic substances," in M.H.B. Hayes, P. MacCarthy, R.L. Malcolm, and R.S. Swift (Eds.) *Humic Substances II. In Search of Structure*, copyright © Wiley, Chichester, UK; b, Reprinted with permission from *Tetrahedron* 23:1653–1667, N.M. Atherton, P.A. Cranwell, A.J. Floyd, and R.D. Haworth, "Humic acid I. ESR spectra of humic acids," Copyright 1967, Pergamon Press PLC.)

III. Organic Free Radicals in Humic Substances

Humic substances (HS) represent the largest portion of the nonliving soil organic matter. They derive from the chemically and microbially operated decomposition and resynthesis of plant and animal decays and are characterized by colloidal, polymeric, polycondensed structures of prevalently aromatic character, of high molecular weight, containing chemically reactive functional groups of various natures, predominantly acidic, that is, carboxylic and phenolic groups. Humic substances are operationally divided into two fractions, humic acid (HA), which precipitate at pH ≤ 2.0 from the alkaline soil extract, and fulvic acid (FA), which are soluble over the entire pH range. FA differs from HA mainly in its lower molecular weight and higher content of reactive oxygenated functional groups (Schnitzer, 1978; Stevenson, 1982).

Both HA and FA are known to be rich in free radicals, which are indigenous to their molecular structure and represent one of their most peculiar reactive properties (Senesi and Steelink, 1989).

A. The ESR Spectrum

In the usual experiment, the ESR spectrum of humic free radicals (HFR) is obtained at RT by an ESR spectrometer operating at X-band frequency with a 100-kHz magnetic field modulation. Solid samples are usually placed in the resonant cavity packed in columns in suitable ESR quartz tubes; for dissolved samples, usually packed in water, a special ESR flat cell is used. The magnetic field is swept over a relatively narrow scan range (generally $\leq$ 100 G), through the field at which the free electron resonates ($g = 2.00232$). Most organic free

Table 1. ESR parameters for some HAs and FAs

Sample	State	Free radical concentration (Spins/g·10^{-17})	Line width (Gauss)	Spectroscopic splitting factor (*g*-value)
Soil HA[a]	Solid	5–10	4.8–5.2	2.0032–2.0047
Soil FA[a]	Solid	1– 2	4.5–7.5	2.0037–2.0050
Soil HA[b]	H_2O solution	2.1	2.5	2.0037
Soil FA[b]	H_2O solution	1.3	2.5	2.0038
Loam soil HA[c]	Solid	1.1	7.1	2.0047
Sludge-amended loam soil HA[c]	Solid	0.4–0.7	6.9–7.5	2.0037–2.0043
Soil fungal HA[d]	Solid	0.6–6.4	4.8–6.6	2.0040–2.0046
Buried soil HA[e]	Solid	7–11	3.9–4.4	2.0045–2.0051
Peat HA[f]	Solution in 0.1*M* NaOH	9.2	2.0	2.0043

Sources: a = Chen et al., 1978; b = Varadachari et al., 1983; c = Senesi et al., 1989b; d = Senesi et al., 1987b; e = Calderoni and Schnitzer, 1984; f = Steelink, 1966.

radicals resonate in fact, at a field corresponding to *g*-values close to this number. Common conditions used for some ESR operating functions in the laboratory of the author are microwave frequency, 9.52 GHz; microwave attenuation, 13 dB, corresponding to a microwave power of about 10 mW, at which the signal is generally least saturated; and modulation amplitude, 6.3 Gauss.

The typical ESR spectra of HFR are featured by a single-line resonance, devoid of any hyperfine structure (Fig. 5a), while a partially resolved hyperfine structure is rarely observed (Fig. 5b).

B. Chemical Nature and Origin

The analysis of the ESR spectrum of HFR provides at most three types of spectral parameters of importance for the characterization of their nature and origin: the *g*-value, the width of the absorption line, and, eventually, the hyperfine structure. ESR data measured for some representative HA and FA originated from soil and related materials are presented in Table 1 and are used in the following discussion.

The spectroscopic splitting factor, that is, the *g*-value, is initially used to characterize the nature of the HFR (Blois et al., 1961; Senesi and Steelink, 1989). It can be approximated from the magnitudes of the magnetic field at which resonance occurs for the sample and for a standard char of known *g*-value, usually N,N-diphenylpycrylhydrazyl (DPPH) diluted in powdered KC1 ($g_{DPPH} = 2.0036$). The *g*-values commonly measured for HA and FA originated from soils and related materials do not differ significantly one from another (see Table 1). They are consistent with semiquinone radical units possibly conjugated to aromatic rings (g = 2.0041 for 9,10-anthraquinone), although a contribution from methoxybenzene radicals (*g*-values range from 2.0035 to 2.0040) and nitrogen-associated radicals (g = 2.0031–2.0037) cannot be excluded (Blois et al., 1961; Senesi and Steelink, 1989). Extended aromatic network conjugated to the semiquinone moiety causes partial delocalization of the free electron from the oxygen atom of the semiquinone to aromatic carbon atoms, with the consequent lowering of *g*-values (Schnitzer and Skinner, 1969; Riffaldi and Schnitzer, 1972a).

The width of the absorption line (line width) is generally measured in Gauss, as the peak-to-peak separation of the first derivative ESR signal. Water solutions of HS always exhibit spectral line widths narrower than those of solid state samples (see Table 1). This behavior may be ascribed to rapid tumbling of molecules in liquids, great freedom of rotation and low association (low hydrogen bonding) with neighboring molecules. Fulvic acids usually show line widths slightly greater than HA of the same source (see Table 1). Small variations in the line widths of ESR signals may be ascribed to a different number of superimposed resonances at slightly differing field values, all contributing to the signal. Line broadening may thus reflect a partial delocalization of the unpaired electron from the semiquinone onto the conjugated aromatic network. Broadening of the line width can also be accounted for by unresolved superhyperfine interactions of the free electron with a number of neighboring aromatic and aliphatic hydrogen nuclei. Other factors that can influence the line width are the free radical concentration of the sample, its state of aggregation (solid state or solution), temperature, solvent–solute interactions, interactions with metal ions, and power-saturation effects.

Unfortunately, in most cases the ESR spectra of HFR lacking hyperfine structure (see Fig. 5a), which renders impossible any further description of the chemical and structural environment around the radical itself. Some degree of hyperfine structure was found, however, in the ESR spectra of soil HS by some authors. A three-line spectrum, which was rationalized in terms of two separate radical species—a semiquinone-type radical and a quinhydrone-type radical—was reported (Steelink and Tollin, 1962). Four-line hyperfine structures (see Fig. 5b) attributed to the interaction of the unpaired electron with two nonequivalent hydrogen nuclei, were observed for a number of acid-boiled HA originated from acid soils and peats (Atherton et al., 1967; Cheshire and Cranwell, 1972). ESR results suggested that the type of HFR in soil HA is highly related to conditions of pH attending the HA formation. A triplet spectrum observed for an oxidized soil FA was ascribed to the interaction of the unpaired electron on the oxygen of a semiquinone unit with two adjacent equivalent hydrogen nuclei (Senesi et al., 1977a). ESR spectra of 3,4-dihydroxybenzoic acid semiquinone anion, of some flavonoids and quercetin and rutin, both of which possess a 3,4-dihydroxyphenyl structural unit obtained after standing for few hours in alkaline solution, resemble that of HFR shown in Figure 5b (Steelink, 1987). A relation was found between dihydroxybenzoic acid groups in HS and their spin content (Schnitzer and Levesque, 1979). These findings suggest that HFR may be at least partly associated with a catechol moiety.

C. Concentration

The concentration of unpaired electrons (free radicals) is probably the most important piece of data that can be obtained from the ESR spectrum of HFR and the most frequently cited ESR parameter, but it is probably also the most difficult and least accurate of ESR measurements. Absolute measurements of spin concentration require knowledge of a large number of factors, all of which may be considerably in error or may vary during the experiment. Spin concentration is, therefore, usually estimated by comparing the ESR signal of the sample with that of a standard char containing a known number of paramagnetic centers. Since the area under an ESR absorption curve is directly proportional to the number of paramagnetic centers contributing to the resonance, it is relatively simple to determine the spin concentration (spins/g) of the sample by comparing its signal area with that of a standard. Because the spectrometer output is in the form of a first derivative curve, a double integration should be performed. Since the line shapes of the sample and standard are identical (Lorentzian shape), the comparison can be made by assuming that the area is proportional to the product of the height and square of the width of the first derivative signal.

Because the measurement of the spin content depends on a comparison with a standard sample, this requires that the sample be measured in an environment identical to the standard, namely, equal volumes of solid sample and standard should be measured simultaneously in the cavity. In most cases this is impossible; thus, some operators use a double cavity instrument while others use concentric sample tubes. Most operators use a single sample tube and attempt to place the standard tube in exactly the same position in the resonant cavity. In the case of solution samples, the choice of solvent may affect the absorption of microwave power. The pH and the time also strongly affect the free radical content measured (Senesi and Schnitzer, 1977). Other factors that affect precision and accuracy in the spin content measurement include power saturation, as previously discussed in this chapter, and the shape of the curves. For a meaningful comparison of signal areas, it is required that the line widths and the spin concentration of standard and sample be comparable, and the same spectrometer settings be used.

Despite these limitations, where a single sample is subject to a sequence of external changes, such as time, pH, irradiation, chemical treatment, and so on, a reasonable precision and reliability in the relative changes measured for spin concentrations and other ESR parameters are expected.

Data in Table 1 show that FA usually has one-third to one-fifth the spin content of HA from the same source. HA isolated from a soil amended with composted urban sewage sludges at different rates and for various times in any case shows a decrease in HFR content compared to that of the native soil HA (Senesi et al., 1989a). Humic acid–like polymers synthesized by various soil fungi show HFR concentrations ranging within 1 order of magnitude, and generally lower than those measured in bulk soil HA. Buried soil (paleosol) HA generally shows HFR content to a little higher than agricultural soil HA. HA from peat also appears to be relatively rich in HFR. Although some general relationship have been suggested to occur between soil type and spin content of HA (Riffaldi and Schnitzer, 1972a; Chen et al., 1978), no specific correlation has been validated.

Soil organic matter fractions isolated from a forest and a marsh soil by a sequential extraction procedure employing mild organic solvents of increasing polarity gave ESR spectra characterized by the typical resonance line of the semiquinone free radical, whose intensity tends to increase with increasing polarity of the extractant (Senesi et al., 1983). The tetrahydrofuran extract resulted the richest in free radicals of both soils. The ESR spectra of some fractions were also characterized by the presence of an additional signal tentatively assigned to a sulfur free radical.

Although the absolute amount of unpaired electrons per mole of HS generally appears to be small, it may contribute significantly to chemical reactivity and functions exerted by HS. For example, a concentration ranging between 10^{17} and 10^{18} spins/g in compounds of molecular weight 10,000 dalton (i.e., HA) represents about one radical per 600 to 60 molecules. In the case of higher molecular weight HS, that is, with MW = 100,000 dalton, however, the value increases considerably to one radical over 60 to 6 molecules.

D. Factors Affecting Concentration

The HFR concentration has been shown to depend on numerous environmental and laboratory factors, such as pH, redox conditions, irradiation, acid-hydrolysis, methylation, and temperature.

Alkaline solutions as well as solid sodium salts of HS are characterized by HFR concentrations that are substantially greater than those of neutral or acidic systems (Steelink, 1966; Wilson and Weber, 1977; Senesi and Schnitzer, 1977, 1978; Ghosh and Schnitzer, 1980). The elevated HFR concentrations measured at alkaline pH are not time-stable, however, but they decrease, first rapidly (within 4 to 5 hours) and then slowly (over the next 6 days) (Senesi and Schnitzer, 1977, 1978). A direct quantitative relationship was developed between pH and HFR content of FA by Wilson and Weber (1977), who concluded that a group of closely related semiquinone species, and not a single entity, are responsible for HFR content variation with pH. These results may be explained by the shifting of the fundamental equilibrium involving the quinone–hydroquinone (i.e., quinhydrone) electron donor–acceptor system, indigenous to HS and generally considered the principal responsible of the generation and maintenance of semiquinone radicals in HS, toward the formation of the stable, long-lived radical anion species, according to

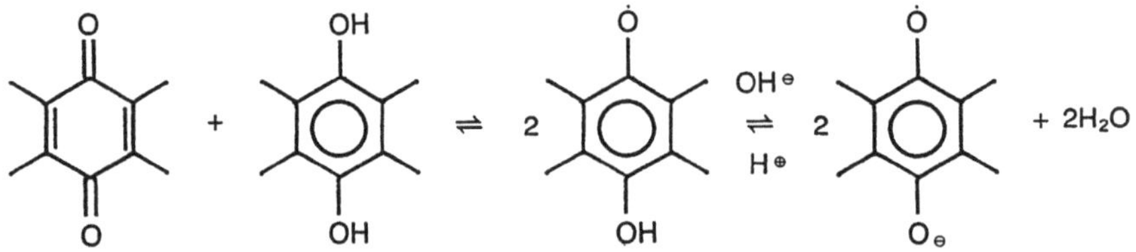

(quinone unit) hydroquinone unit) (semiquinone radical) (semiquinone radical anion) (12)

At high pH values, the concentration of the radical may, therefore, approach that of one of the two partners in the parent donor–acceptor system. Hydroxy–

anthraquinones, tannins, and lignins, all containing quinone and phenolic groups—that is, potential electron donor–acceptor systems—also showed similar behavior (Steelink and Tollin, 1966). For instance, the hydroxyanthraquinone alizarin was shown to contain one free radical per 1,000 molecules in the free state, whereas the solution at pH 14 contained one free radical per 4 molecules (Steelink and Tollin, 1966), A quinone content of 0.5 mmole/g observed for some HA (Schnitzer and Riffaldi, 1972) would theoretically yield 6×10^{20} spins/g of semiquinone anion in the presence of a base and a suitable donor group. Values approaching this number have been observed in some humate salts (Tollin et al., 1963; Steelink, 1964, 1966).

Considerable evidence exists of a rapid (10 to 40 min) and marked (10- to 100-fold) increase of HFR content in HA and FA solutions at various pH values, upon chemical reduction (Tollin et al., 1963; Senesi and Schnitzer, 1977, 1978), although some contrasting results are reported (Steelink, 1964; Atherton et al., 1967). Gradual decreases in spin concentration were measured in FA solutions at any pH after maximum values were attained.

Although enhanced ESR signal intensity was observed in some cases with exposure to air or high-temperature oxidation of alkaline solutions of HA (Steelink, 1964; Atherton et al., 1967; Riffaldi and Schnitzer, 1972a), chemical (oxygen or other inorganic oxidants) or electrochemical oxidation was shown generally to produce a time- and pH-dependent decrease of HFR content of freshly prepared HA or FA solutions (Slawinska and Slawinski, 1975a; Wilson and Weber, 1977; Senesi and Schnitzer, 1977; Senesi et al., 1977b). The highest decay rate was observed at the highest pH value, while minimum values were attained in any case within 4 hours (Senesi and Schnitzer, 1977; Senesi et al., 1977b).

Effects of oxidation and reduction on HFR concentration of FA solutions at various pH values are apparently reversible (Senesi et al., 1977b; Senesi and Schnitzer, 1978). This behavior is accounted for by the reversible reaction schemes outlined here, occurring at acidic or neutral pH (eq. 13) and alkaline pH (eq. 14), respectively:

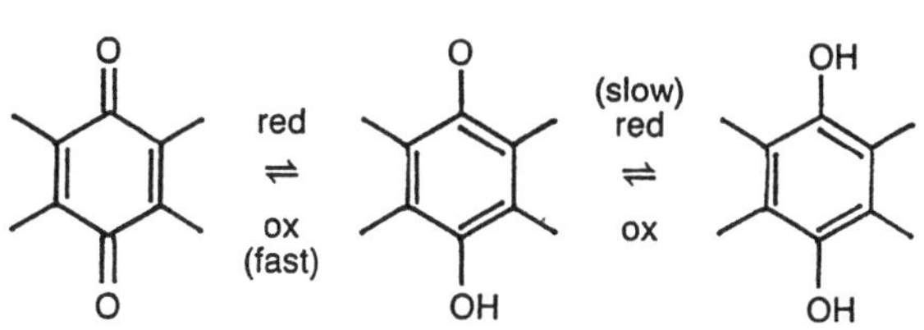

(13)

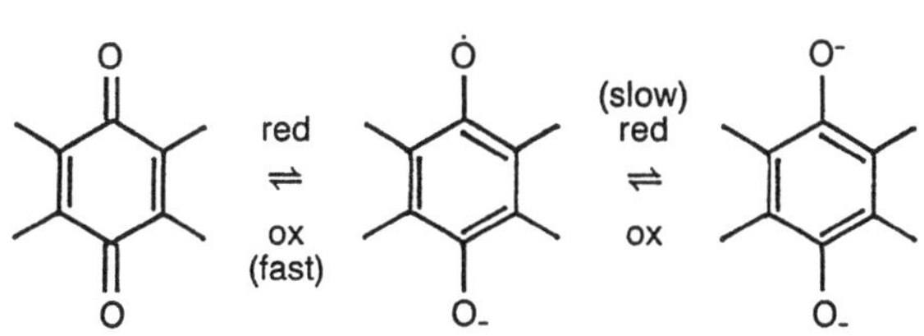

(14)

Visible- and ultraviolet (UV)-light irradiation of HA and FA in either the solid state or water solution at various pH values reversibly enhances the ESR signal intensity (Lagerkrantz and Yhland, 1963; Slawinska et al., 1975, 1978; Slawinska and Slawinski, 1975b; Senesi et al., 1977b, 1977c; Senesi and Schnitzer, 1977, 1978). In FA solutions, maximal values of HFR concentration were attained generally within 5 minutes of irradiation, while the HFR content returned rapidly to a value similar to those before irradiation when the light was switched off (Senesi and Schnitzer, 1977, 1978). Vis-light of wavelengths ranging from 550 to 600 nm (green–yellow–orange) seemed the most efficient in increasing HFR concentration, especially in FA solutions at alkaline pH (Senesi and Schnitzer 1977). Irradiation of previously oxidized FA solutions produced an increase of HFR concentration, especially at high pH, while no effect was measured on FA solutions that had attained maximum HFR concentration on chemical reduction (Senesi et al., 1977b; Senesi and Schnitzer, 1978). Photoinduced, charge-transfer mechanisms occurring between donor groups containing mobile hydrogen atoms and acceptor dark pigments, both present in HA, were suggested as the main factors responsible for the increase of HFR content upon irradiation (Lagerkratz and Yhland, 1963). On the other hand, semiquinones are known to be produced by photoirradiation of quinones in solution, in the presence of proton donors (Bruce, 1974). Also suggested was the hypothesis that light irradiation might promote the spin-coupled system HA–O_2 into electronically excited states with high chemical reactivity, thus yielding photodegradation products rich in free radicals, probably formed in an exothermic oxidative ring cleavage of quinone rearrangement in HA. Humic substances might therefore be expected to act as photosensitizers in the light- and O_2-stimulated detoxication processes that HA-bonded chemicals, such as herbicides, might undergo in soil (Senesi and Chen, 1989). Tichy (1971) observed an increased biological activity of HA enriched in free radicals by UV irradiation. Khan and Gamble (1983) found that irradiation of HA in the presence of s-triazine derivatives caused photodegradation of the s-triazine.

In conclusion, pH-raising, reductive conditions, and light irradiation have been shown to enhance the total HFR concentration of HS, especially in water solution, while leaving almost unaltered both the *g*-value and the line width of the ESR signal. These results therefore indicated the production of short-lived, "transient" free radicals of basically the same nature of indigenous, stable HFR, by substantially similar reversible mechanisms based on an electron donor–acceptor (or charge-transfer) model of HA that contains quinone and hydroquinone moieties coexistent with semiquinone radicals.

Acid hydrolysis of HA generates products characterized by a free radical content increasing as the period of hydrolysis increases (Steelink, 1964; Steelink and Tollin, 1967; Atherton et al., 1967; Riffaldi and Schnitzer, 1972b; Saiz Jimenez and Shafizadeh, 1985). A more aromatic structure enriched in free radicals appeared to result from the removal of the nonaromatic, hydrolyzable portion, such as carbohydrates and peptides, of the humic macromolecule and the probable condensation and polymerization of hydrolysis-resistant structural units.

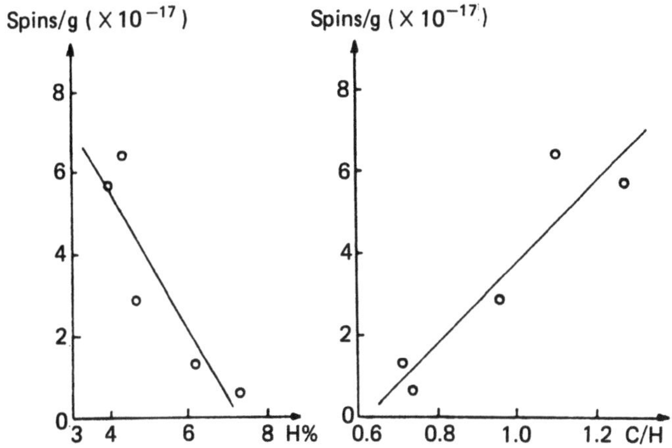

Figure 6. Plot of H% and C/H atomic ratio of humiclike soil fungal polymers versus free radical concentrations (Senesi et al., 1987b).

Although contradictory results are reported on the effect of methylation, the tenfold decrease of HFR concentrations measured for podzolic HA and FA seems to confirm the OH groups as the most important electron donors responsible for the formation and existence of HFR (Schnitzer and Skinner, 1969).

Exposure of solid HA to increased temperatures up to 450°C caused a 100-fold increase in HFR content, probably as a result of homolytic bond-cleavage and depolymerization reactions (Steelink and Tollin, 1962; Schnitzer and Skinner, 1969; Bel'kevic et al., 1974).

High-energy (electron and gamma) irradiation of HA and FA, in either the solid state or aqueous solution at various pH values, produces variations of the HFR concentration probably related to changes in molecular dimension induced by either polymerization–aggregation or depolymerization–dispersion phenomena occurring in HA and FA upon irradiation (Senesi et al., 1977d).

In conclusion, application of ESR spectroscopy contributes to the elucidation of mechanisms of important natural chemical, photochemical, and biological processes involving HS and the reactive species they produced, which occur either during the genesis of HS (e.g., temperature raising and natural soil radioactivity), or in the actual soil–water–organisms system under various conditions of pH, redox potential, sunlight irradiation, and hydrolysis, possibly involving organic chemicals, enzymes, and metal ions.

E. Structural and Reactive Implications

Concentration of HFR in HS of various natures and origins was found to correlate positively to the E_4/E_6 ratio, absorbance at 465 nm, atomic C/H ratio, and 0% and negatively to H% (Fig. 6; Schnitzer and Skinner, 1969; Riffaldi and Schnitzer, 1972a; Chen et al., 1977; Retcofsky et al., 1981; Stuermer et al.,

1978; Senesi et al., 1987b). This was taken as an evidence that HFR content is directly related to the dark color and degree of aromaticity, molecular complexity, and particle size of HS. A good correlation also appeared to exist between ESR and infrared (IR) data. The predominant presence of IR absorptions attributed to aromatic structures is generally associated with high HFR content in HA, while low HFR contents correspond to a prevalence of aliphatic and olephinic absorptions in the IR (Senesi et al., 1987b).

With the major role played by HFR in polymerization–depolymerization reactions involved in the synthesis and transformations of HS in soils, ESR spectroscopy has found some useful applications in studies related to humification and mineralization phenomena. Free radical concentration of 0.5-M extracts of peat samples was found to correlate significantly either with the pyrophosphate index, that is, the optical density measured at 550 nm for dilute pyrophosphate solution of peat—generally used to estimate their humification degree—or with the concentration of phenolics in the extracts (Schnitzer and Levesque, 1979). The free radicals involved in peat humification processes appeared to be substituted semiquinones, which were present in increasing concentrations as the humification proceeded. The authors reported their conclusions as preliminary and recommended the ESR method worth further experimentation, since it appeared the simplest and least time-consuming procedure so far proposed for assessing humification in peat.

Two recent reports (Eloranta et al., 1980; Eloranta and Hannonen, 1985) have shown that the free radical concentration, as well as the ferric complex–ESR signal in peat, increased with increasing degree of humification. Very recently, Wikander and Nordén (1988) found that ESR signal intensity in both free radical and high spin–ferric complexes of a number of peat samples varied according to their botanical class and their degree of decomposition. Only for a well-defined class of *Sphagnum* peat, a quantitative relationship was found between ESR parameters and the degree of humification, while general good correlations existed between ESR parameters and amino acid constituents.

Addition of Fe^{3+} ions to solutions of HA had the effect of decreasing HFR content and inhibiting further polymerization (Chukov and Nikonova, 1980). It was suggested that these results relate to the low degree of polymerization and the low molecular weights of HA in soils with high Fe^{3+} contents and also explain the low HFR content measured in HA from the Fe^{3+}-rich B_h horizon of a podsol soil, with respect to the high HFR content of HA from the A horizon of the same soil. ESR data supported the conclusion that recently formed, low-molecular-weight HS did not undergo further polymerization processes because their free radicals combined with Fe^{3+} ions and moved to the B_h horizon (Chukov and Gurov, 1984). ESR spectroscopy revealed that the iron extracted from the B_h horizons of iron humus podzols by EDTA at pH 9.1 was predominantly in the form of polymeric Fe^{3+} hydroxyde and oxide associated with organic matter (McBride et al., 1983). Also furnished was ESR evidence on the prevalent involvement of aromatic HFR in reactions with metal ions at high cation (Fe^{3+} or Cu^{2+}) to HA ratios, while at low metal concentrations HFR associated with aliphatic humic components reacted prevalently (Chukov et al., 1983). The regular increase of HFR signal and the simultaneous decrease of Fe^{3+} ESR signal observed during the stepwise purification of the original HA, indicated that a reversible restoration of HFR took place when HA was freed from cations, particularly iron (Chukov and Tsciplenkov, 1983).

On the basis of ESR data, Suflita et al. (1981) showed that a soil-extracted organic catalyst capable of mediating oxidative coupling reactions was able to generate hydroxy radicals when incubated with phenolic substrates. These radicals were suggested to be essential intermediates in the oxidative coupling, that is, condensation, of phenols in the formation of humic polymers.

Wang et al. (1983) have applied ESR spectroscopy in analyzing the role of natural clays, soil oxides, and whole soils as catalysts in the oxidation of phenolic compounds and polymerization of mixtures of phenolic, nitrogenous, and other organic compounds to yield HS. The authors ascertained that ESR spectra of HS synthesized from phenolic mixtures in the presence of these catalysts were similar to those reported for native HS and for model HS prepared by polymerization of simple organic compounds in the presence of known metal ion catalysts.

More recently, the ESR spectrum of semiquinone radical was revealed in the aqueous phase of a reaction mixture of goethite and hydroquinone (McBride, 1987). Single electron transfers able to generate the radical were suggested to occur at the iron oxide and manganese oxide surfaces during the oxidation of hydroquinone.

ESR studies of reactions of benzene and other arenes in Cu^{2+}- and Fe^{3+}-exchanged smectites gave evidence of the generation of two ESR- distinguishable chemical species upon chemisorption of benzene by clay under dehydrating conditions (Mortland and Pinnavaia, 1971; Rupert, 1973; Pinnavaia et al., 1974). Type I species had an ESR spectrum dominated by the resonance produced by the metal ion exchanged into the clay (π-type, metal–benzene complexes), while type II species showed a single-line ESR spectrum apparently consistent with organic free radicals that had less aromatic character than benzene. ESR studies by Rupert (1973) suggest that the reaction forming the type II species involved electron transfer from benzene to Cu(II). More recently, Eastman et al. (1984) showed by ESR that the reaction of benzene with exchangeable Cu(II) and Fe(III) in hectorite clay films determined a decrease in the oxidation state of the metal ions and produced a variety of organic radical products, the nature of which depended on the concentration of water in the reaction medium and the reaction time. These findings may be important in the polymerization and humification processes that occur in natural soil conditions.

IV. Free Radicals in Humic Substance–Herbicide Interactions

Humic substances are generally considered to greatly affect phytotoxicity and bioavailability, degradation and detoxication, accumulation, mobilization and transport, volatility, and leachability of organic chemicals in the soil–water–organism system. Senesi and Chen (1989) recently reviewed the multiple effects that HS may exert by interacting with a variety of organic chemicals, including adsorption, forming chemical bonds of various natures and strengths, partitioning and solubilizing effects, catalytic and hydrolysis effects, enzyme-mediated interactions, and degrees of photosensitization.

In particular, a variety of mechanisms have been proposed for the chemical binding of organic chemicals to HS, including ionic and hydrogen bonding, weak electrostatic and Van der Waals attractions, ligand exchange, and complexation through polyvalent metal ions (Hayes, 1970; Stevenson, 1972; Weed and Weber, 1974; Senesi and Chen, 1989). Active participation of free radicals in some of the interaction mechanisms proposed has also been suggested by some authors (Stevenson, 1972), but application of ESR spectroscopy has only recently provided experimental evidence of their role in binding mechanisms such as charge-transfer (electron donor–acceptor) processes and cross-coupling reactions (Senesi, 1981a; Senesi and Testini, 1982, 1983a,b, 1984; Senesi et al., 1984, 1986a,b, 1987a,e).

A. s-Triazines and Substituted Ureas

Increases of various entities in free radical concentration and broadening of ESR spectral line width were observed, with respect to the values measured for original HA, in the interaction products obtained at RT and under air atmosphere and natural light, between HA from soil and other soil-related sources and diluted water solution of s-triazine or substituted urea herbicides, while *g*-values remained almost unchanged (Senesi, 1981a; Senesi and Testini, 1982, 1983a, 1983b; Senesi et al., 1987e). ESR data of some representative samples are presented in Table 2.

ESR results were explained by assuming that electron-deficient quinonelike structures in the HA molecule induce a single-electron transfer from the electron-rich nitrogen atoms of the herbicide molecule, giving rise to the formation of radical cation and anion species in a charge-transfer complex, according to the following reaction:

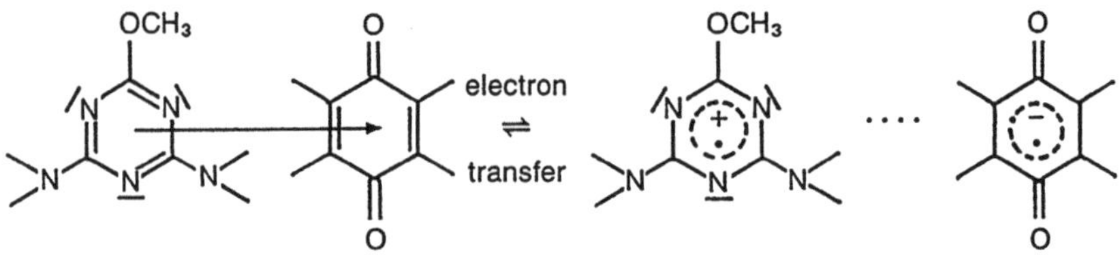

(s-triazine; electron–donor) (humic quinone; electron–acceptor) (radical cation and anion; charge-transfer complex) (15)

A light-induced unpairing of electrons (photoinduced charge-transfer) may also account in part for the increase observed in free radical content in the interacting system, similarly to what occurs for several different model and natural organic

Table 2. Electron spin resonance spectral data for humic acids from soil and related materials, and their products of interaction with some herbicides representative of the classes of s-triazines (prometone), substituted ureas (fenuron), and chlorophenoxyalanoic compounds (2,4-D and 2,4,5-T)

ESR parameters	Free radical concentrations (Spins/g × 10^{-17})					Line widths (Gauss)					Spectroscopic splitting factors (*g*-values)				
Humic acid / Herbicide	Soil	Synth.[a]	Peat	Coal	Comp.[b]	Soil	Synth.	Peat	Coal	Comp.	Soil	Synth.	Peat	Coal	Comp.
Prometone	23.56	449.90	4.36	2.75	4.76	8.2	4.4	7.1	6.2	5.7	2.0035	2.0035	2.0037	2.0036	2.0037
Fenuron	33.87	355.54	3.09	2.43	6.32	6.5	4.9	7.2	6.4	5.9	2.0033	2.0034	2.0036	2.0037	2.0037
2,4-D	5.52	30.29	0.54	1.56	0.62	6.5	5.4	7.0	7.2	6.3	2.0034	2.0037	2.0039	2.0039	2.0038
2,4,5-T	5.54	26.29	0.31	1.28	0.90	6.6	5.3	7.1	7.0	6.3	2.0032	2.0041	2.0037	2.0035	2.0036
Original humic acid	8.62	74.84	1.50	1.89	2.90	6.2	4.1	5.4	6.2	6.2	2.0035	2.0032	2.0033	2.0035	2.0037

Source: Senesi and Testini, 1983a; Senesi et al., 1984, 1987a.

[a] Synthetic.

[b] Compost.

donor–acceptor systems (Foster, 1969; Lagerkrantz and Yhland, 1962). Further, unpaired electrons are highly stabilized in systems of increased molecular complexity, deriving from the covalent binding which may occur between quinones and other carbonyl groups of HA with amino groups of the herbicide, through condensation and nucleophilic addition reactions (Hayes et al., 1975; Parris, 1980).

An inverse relation exists between phenolic OH or carboxylic functional group content of HA and free radical concentrations measured in their interaction products with s-triazines (Senesi and Testini, 1982). This result indicates that, with decreased capacity of HA to form ionic and hydrogen bonds with the weakly basic s-triazines, their tendency to interact with these herbicides through electron donor–acceptor mechanisms increases. In this respect, it has been shown that the free radical content of model electron donor–acceptor systems is generally directly related to the tendency of these systems to form and exist (Bijl et al., 1959). A direct proportionality was observed between free radical concentration in HA–urea interaction products and the bioactivity of corresponding ureas, expressed as inhibitory potency of the herbicide in the Hill reaction (pI_{50} index) (Senesi and Testini, 1983b). This suggests that single-electron donor–acceptor mechanisms involving similar free radical species and influenced by similar molecular parameters may be responsible for both the inhibitory activity of these herbicides in plant chloroplasts and their adsorption by HA in soil.

The chemical structure and properties of the herbicide also appeared to affect to some extent the formation of charge-transfer systems with HA. For instance, the highest efficiency of prometone, among the s-triazines, in forming electron donor–acceptor systems with HA (the highest increase in free radical concentration) was attributed to the presence of the electron-donating methoxyl group on the 2-position and the isopropyl group on each amino group in 4- and 6-positions of the triazine ring (Senesi and Testini, 1982; Senesi et al., 1987e). The highest increase of free radical concentration exhibited in the interaction products with HA by fenuron, among substituted ureas, was attributed to the absence of deactivating chlorine atoms on its phenyl ring, which is expected to reinforce the electron donor tendency of fenuron with respect to other ureas that contain one or two chlorine atoms on the ring (Senesi et al., 1987e).

B. Chlorophenoxyalkanoic Compounds

The products of interaction between a number of water-dissolved chlorophenoxyalkonoic acids and esters, including 2,4-D and 2,4,5-T, and HA from soils and other related sources showed a considerable quenching of free radical concentration initially present in untreated HA. Simultaneously, the line width sensibly broadened, while *g*-values remained almost constant (Table 2; Senesi et al., 1984, 1987a).

ESR results agreed well with observations made previously by other authors not using ESR spectroscopy (Crosby, 1976; Stott et al., 1983; Bollag, 1983). The possible covalent incorporation of chlorophenoxy units into HA was thus suggested to occur by homolytic cross-coupling reactions between indigenous HFR

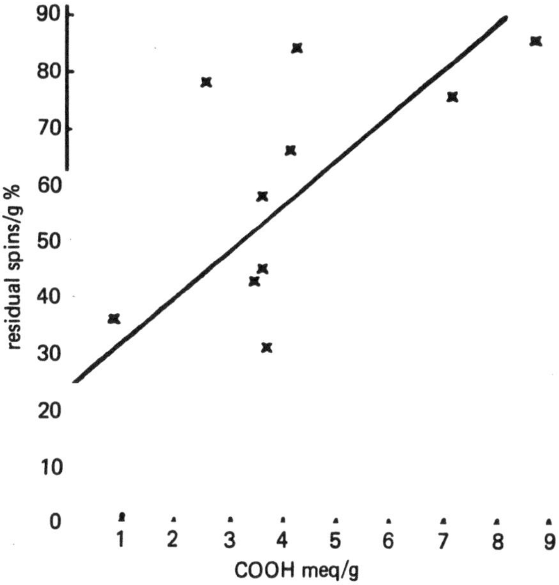

Figure 7. Average residual spin concentration (%) in humic acid–phenoxy compound interaction products versus COOH content of humic acid (Senesi et al., 1986a).

and free radical intermediates generated in the preliminary chemical, photochemical, and/or biological degradation of the herbicide molecule, according to the following reaction:

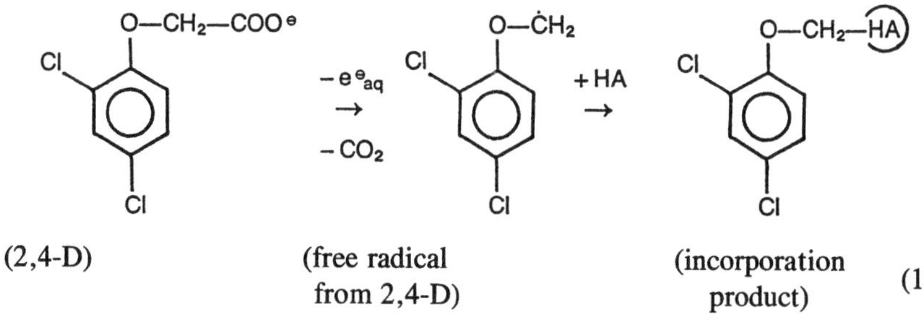

(16)

The nature and origin of HA appeared to influence their reactivity toward chlorophenoxy herbicides. In particular, the carboxyl content or the COOH to phenolic OH ratio of HA correlated negatively to the residual free radical concentration, which is assumed to be inversely proportional to the interacting capacity of the system (Fig. 7; Senesi et al., 1986a, 1987d). The molecular structure of the chlorophenoxy unit also affected the reactivity for coupling with HA. Most likely, chlorine atoms on the phenoxy ring interfere to a certain extent with cross-linking to the humic macromolecule (Senesi et al., 1986a).

C. Amitrole, Alachlor, and Cycloate

ESR spectra of interaction products of amitrole, alachlor, and cycloate with HA from soil and other sources showed a decrease in free radical concentration in the soil HA, while the opposite was true for HA from other sources (Senesi et al., 1986b). A specific, differentiated role played by free radicals in these interactions was suggested, but further work was considered necessary to explain the mechanisms involved.

V. Metal–Humic Substance Complexes

Many important phenomena in which metals are involved in the soil–water–plant system, including adsorption–desorption, movement and transport, bioavailability and phytotoxicity, depend critically on their interactions with HS. ESR technique has been widely applied in the study of both naturally occurring and laboratory-prepared model complexes formed by HA and FA of different origins and natures with paramagnetic transition metal ions of great chemical and biological importance to agriculture and environment, including iron, copper, manganese, vanadium, and molybdenum (Senesi and Steelink, 1990). ESR analysis of metal–HS associations can provide useful and, in some cases, unique information on metal oxidation states, mechanisms of metal binding, symmetry and type of coordination sites, identity of ligand atoms and groups, and stability of metal–HS complexes.

The discussion that follows is not intended to be a comprehensive review of the subject but to provide the reader with the principal structural and chemical information that the direct analysis and comparative interpretation of ESR spectra and parameters derived therefrom can furnish about paramagnetic metal–HS complexes. For brevity, the presentation focuses on some representative results obtained by the author and collaborators, while referring appropriately to the vast literature on the topic.

A. Iron Complexes

An example of a relatively simple ESR spectrum of HS, obtained at RT over a field scan range of 10,000 Gauss on a powdered sample of a podzol FA, is shown in Figure 8 (Senesi et al., 1977e). Three ESR signals are apparent. The sharp and intense line centered at $g = 2.00$, always present in ESR spectra of HS and arising from HFR species, was discussed previously in this chapter. The asymmetrical line centered at $g = 4.2$, commonly observed in ESR spectra of HS, is consistent with Fe^{3+} ions bound to humic functional groups in tetrahedral or octahedral sites with highly rhombic symmetry (Senesi, 1981b; Senesi et al., 1977e, 1986c, 1987c; Cheshire et al., 1977; Eltantawy and Baverez, 1978; McBride, 1978; Abdul-Halim et al., 1981; Schnitzer and Ghosh, 1982; Filip et al., 1985; Kallianou and Yassoglou, 1985). This form of iron exhibited considerable resistance to proton and metal exchange and to reduction (Senesi et al., 1977e), thus suggesting that Fe^{3+} ions are strongly bound and protected by HS in inner-sphere complexes. The broad

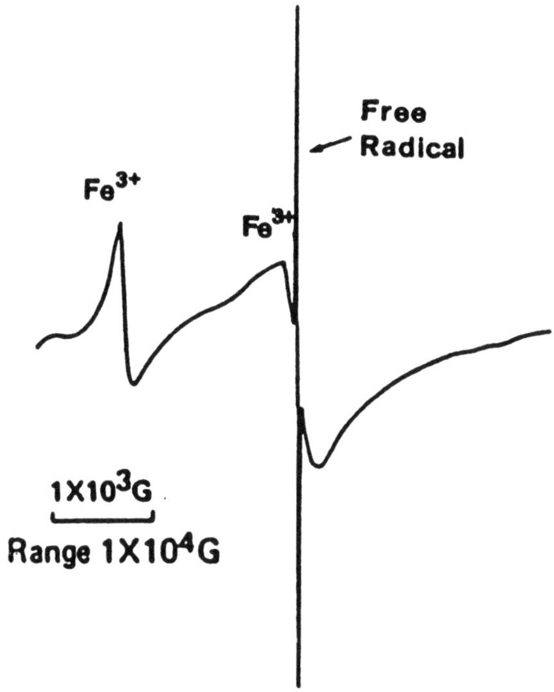

Figure 8. ESR spectrum recorded at RT of a podzol powdered fulvic acid. (Reprinted with permission from *Geochim. Cosmochim. Acta* 41:969–976, N. Senesi, S.M. Griffith, M. Schnitzer, and M.G. Townsend, "Binding of Fe^{3+} by humic materials," Copyright 1977, Pergamon Press PLC.)

signal centered at $g = 2$ probably consists of several superimposed lines arising from extended spin–spin couplings between various paramagnetic metal ions absorbing in this region. Most likely, this resonance is derived prevalently from spin coupling of neighboring high-spin Fe^{3+} ions, loosely held by the humic molecule in octahedral sites with small axial distortion from cubic symmetry (Cheshire et al., 1977; Senesi et al., 1977e; McBride, 1978; Abdul-Halim et al., 1981).

A representative example of a more informative, typical ESR spectrum of natural HS, obtained over a scan range of 8000 G at 77 K on a powdered sample of a soil fungal (*E. echinulatum*) HA is shown in Figure 9. Besides the previously described ESR signals of Fe^{3+} at $g = 4.2$ and HFR at $g = 2.00$, the spectrum features two weak resonances at low field values ($g = 8.8$ and $g = 5.8$), sometimes observed in ESR spectra of HS, the former generally attributed to Fe^{3+} ions in sites with near-orthorhombic symmetry and the latter to high-spin Fe^{3+} ions in strong axially symmetric crystal fields (Cheshire et al., 1977, 1979; Abdul-Halim et al., 1981; Senesi et al., 1985c 1987c). Filip et al. (1985) suggested that the Fe^{3+} ions responsible for the latter resonance may be bound to four nitrogen atoms in a planar configuration possibly deriving from a residual chlorophyll structure

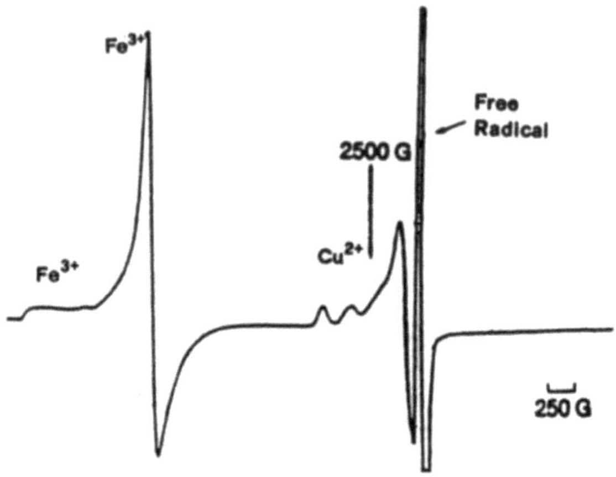

Figure 9. ESR spectrum at 77 K, scan range 8000 G, of a soil fungal humiclike polymer (Senesi et al., 1987c).

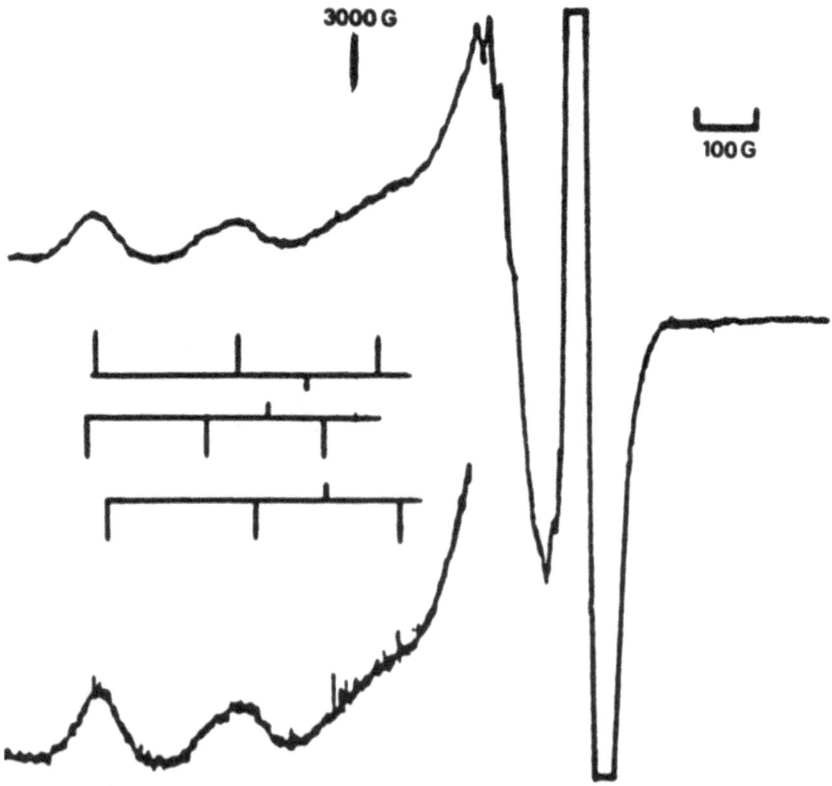

Figure 10. ESR spectrum at 77 K, scan range 2000 G of a soil fungal humiclike polymer; *below* spectrum recorded at a higher gain (Senesi et al., 1987c).

Table 3. ESR spectral parameters for natural Cu^{2+} complexes in humic substances of different origins and nature

Sample origin	State	$g_{\parallel}$	$\lvert A_{\parallel}\rvert$ (10^{-4} cm^{-1})	$g_{\perp}$	Ligand atoms
Soil HA[1]	Solid	2.269	180	2.070	2 O, 2 N
Soil fungal HA[2]	Solid	2.242[a]	190	2.058	1 O, 3 N
		2.283	164		3 O, 1 N
		2.226	196		1 O, 3 N
Buried soil HA[3]	Solid	2.282	176	2.063	3 O, 1 N
Soil FA[4]	Frozen	2.344[a]	158	2.074	4 O
	Solution	2.291	180		3 O, 1 N
Sludge FA[4]	Frozen	2.297[a]	184	2.067	3 O, 1 N
	Solution	2.314	182		4 O
		2.270	187		2 O, 2 N

Sources: 1 = Senesi et al., 1989b; 2 = Senesi et al., 1987c; 3 = Senesi and Calderoni, 1988; 4 = Senesi and Sposito, 1984.

[a] Main resonance.

incorporated in the HS macromolecule, particularly suited to the geometric and chemical features of Fe^{3+} ion which has replaced Mg^{2+} ion.

B. Copper Complexes

Another typical feature of the spectrum in Figure 9 is a rigid-limit pattern in the $g = 2$ region, which can be better interpreted by the analysis of the enlarged spectrum of this region (scan range 2000 G; Fig. 10). The pattern is anisotropic and of the "axial" type, showing a major unresolved absorption at higher field ($g_{\perp}$) associated with a lesser absorption at lower field ($g_{\parallel}$) resolved into a quadruplet, which corresponds to the principal symmetry axes of the molecule lying either perpendicular or parallel to the external magnetic field.

This type of ESR spectrum, often observed in HS of various origins and natures, is evidence of a $d_{x^2-y^2}$ ground state for Cu^{2+} ions held in inner-sphere complexes in HS, with ligands arranged in a square planar (distorted octahedral) coordination around the central ion (tetragonal symmetry; Goodman and Cheshire, 1973, 1976; Cheshire et al., 1977; Lakatos et al., 1977; McBride, 1978; Bresnahan et al., 1978; Abdul-Halim et al., 1981; Boyd et al., 1981, 1983; Schnitzer and Ghosh, 1982; Saiz Jimenez and Shafizadeh, 1984; Senesi and Sposito, 1984; Senesi et al., 1985c, 1986c, 1987c,d; Filip et al., 1985).

Since the nuclear spin of both copper isotopes ^{63}Cu and ^{65}Cu has a value of I = 3/2, the ESR spectrum should be split into four $(2I + 1)$ features both at $g_{\parallel}$ and at $g_{\perp}$, but only the resonance at $g_{\parallel}$ is generally resolved into a quadruplet, while the resonance at $g_{\perp}$ shows splittings only rarely. A more careful analysis of the pattern at $g_{\parallel}$ showed that it consists of three superimposed quadruplets, each component corresponding to a different type of local environment for Cu^{2+} ions bound to HS.

The ESR parameters of Cu^{2+} complexes calculated from the spectra of some humic materials from soils and related sources are listed in Table 3. The g- and A-

values suggest a high covalent bond contribution for Cu^{2+} in HS and are consistent with principal binding sites involving the ligand atoms indicated in Table 3. Carboxyls, phenolic hydroxyls, carbonyls, and eventually water molecules are generally indicated as the main oxygenated functional groups participating in the Cu^{2+} binding by soil HS (McBride, 1978; Bresnahan et al., 1978; Boyd et al., 1981, 1983; Schnitzer and Ghosh, 1982; Senesi and Sposito, 1984; Senesi et al., 1985c, 1987d). The participation of nitrogen in the coordination of Cu^{2+} was also supported by the resolved pattern observed in some cases at $g_{\perp}$, deriving from superhyperfine coupling of the unpaired electron of copper to one or more nitrogen ligand nuclei ($I = 1$). This was accounted for by the proteinaceous components of which some HS, such as sewage sludge FA and fungal HA, are rich (Senesi and Sposito, 1984; Senesi et al., 1987c). A generalized model for a Cu^{2+} complex in HS is shown as follows:

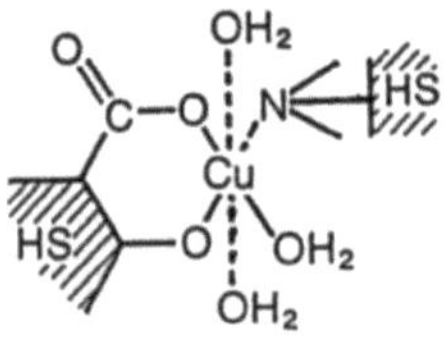

(17)

Additional chemical and structural information can be obtained in some cases from ESR spectra. For example, the ESR parameters of the main resonance of HA synthesized by the soil fungus *E. echinulatum* agree well with those typical of Cu^{2+}-laccase complexes. This suggests the apparent presence of a humus–laccase complex, or the incorporation of Cu^{2+}-containing enzyme fragments, in the humic polymer produced by the fungus (Senesi et al., 1987c). *E. echinulatum* is known, in fact, to excrete laccase for the oxidation and polymerization of phenol units during the synthesis of the fungal polymer (Saiz Jimenez, 1975).

Analysis of the very structured pattern at $g_{\perp}$ shown in the ESR spectra of HA isolated from some peat and mineral soils indicates the existence of Cu^{2+}-tetraporphyrin complexes highly resistant to acid washing (Goodman and Cheshire, 1976; Cheshire et al., 1977; Abdul-Halim et al., 1981). Biologically very stable porphyrin structures, originating from plant chlorophyll and incorporated into the HS, therefore may be deeply involved in copper fixation by HS in peat and soil.

The existence of two classes of binding sites of different strengths for Cu^{2+} ions in soil and aquatic FA, suggested by titration experiments, was confirmed by ESR spectroscopy of frozen solutions (77 K) of Cu^{2+}-FA complexes prepared at various Cu-to-FA ratios (Bresnahan et al., 1978). At low metal-to-FA ratio, ESR data indicate that sites of high binding strength coordinate the Cu^{2+} ion, with four (or three) oxygen donor atoms of carboxylate, phenolate, and carbonyl groups of the same or different FA molecules. At large Cu-to-FA ratios, however, numerous weaker sites predominate and, because water is highly competitive as a ligand in these conditions, possibly only two FA donor atoms are bonded to each Cu^{2+} ion.

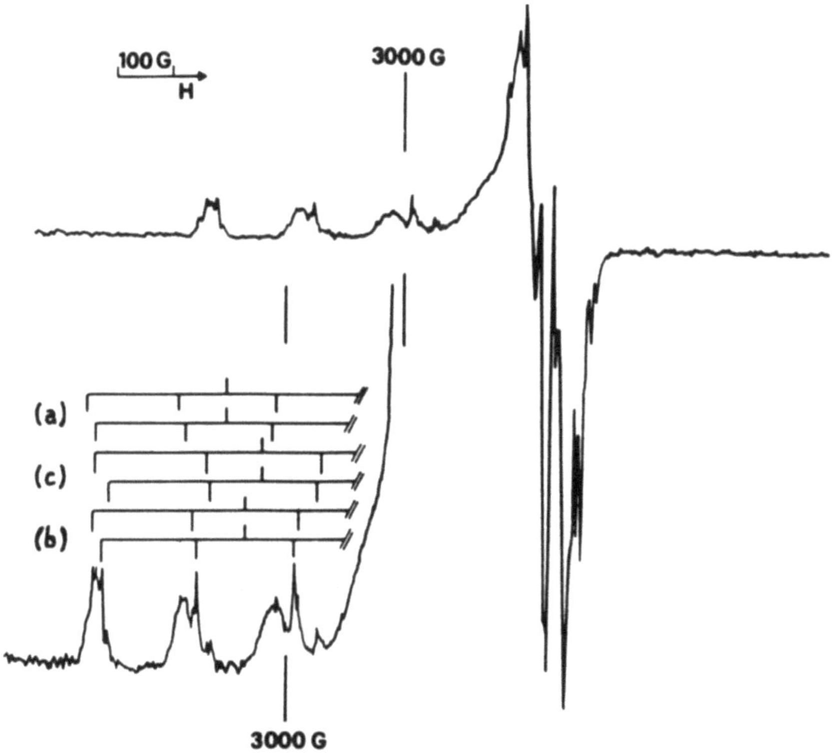

Figure 11. ESR spectrum of Cu^{2+}-sludge fulvic acid in frozen (77 K), resin-treated aqueous solution, at a Cu/fulvic acid ratio of 0.8/1 (Senesi et al., 1985b).

Significant similarities were found between ESR spectra and parameters of laboratory-prepared Cu^{2+}-sludge FA complexes and those of model Cu^{2+} complexes formed by anionic surfactants identified in the sludge FA (Senesi and Sposito, 1987b). ESR data showed a higher speciation of binding sites for copper in FA complexes than in surfactant complexes. Partially degraded surfactant components, with COOH groups at the end of the alkyl chain, may thus act as important ligands to Cu^{2+} ions.

The ESR spectrum shown in Figure 11 is a typical example of the possibility to obtain, for some model Cu^{2+}–humic complexes, highly resolved spectra that may be subjected to a very accurate analysis and interpretation, able to furnish invaluable information on the structure and geometry of Cu^{2+}–HS complexes. The spectrum was measured at 77 K over a scan range of 2000 G on a frozen (77 K) resin-exchanged solution of Cu^{2+}–sludge FA complexes prepared at a Cu-to-FA molar ratio of 0.8:1 (Senesi et al., 1985b). Three different components are exhibited, each resolved in a quadruplet at $g_{\parallel}$, showing couples of closely spaced peaks attributed to hyperfine coupling of the unpaired electron to the two copper isotopes nuclei ^{63}Cu and ^{65}Cu. The ESR spectral parameters calculated for the two

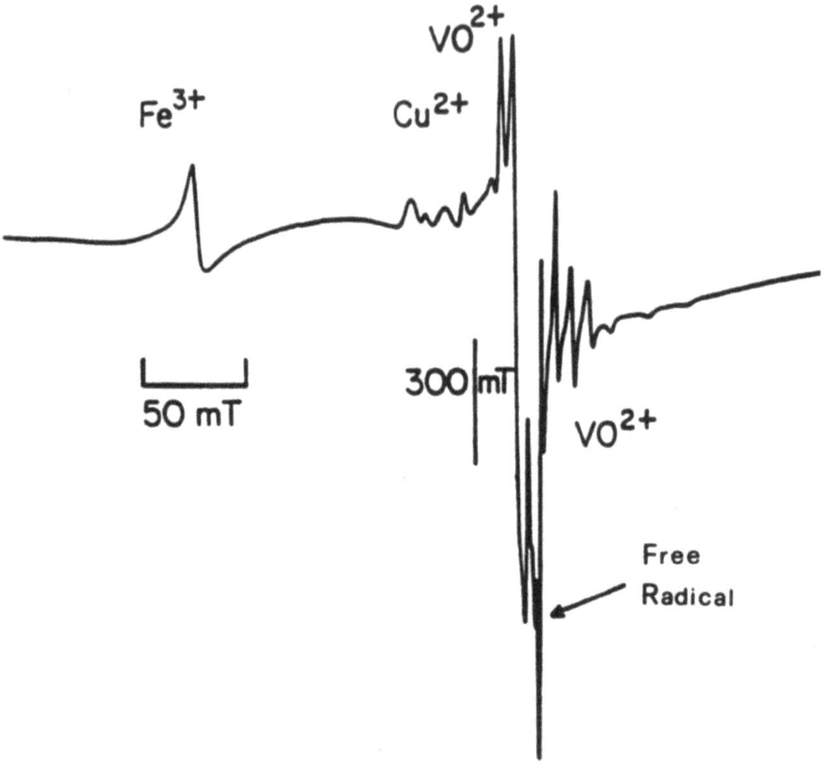

Figure 12. ESR spectrum of a loam soil humic acid powder at 77 K, scan range 8000 G (Senesi et al., 1989b).

prevailing components are consistent with 4 O and 2 O, 2 N coordinating sites, while those of the minor component indicate a more nitrogenated (1 O, 3 N) site involved by the FA in Cu^{2+} binding. The intensely structured pattern at $g_{\perp}$ was tentatively resolved into two groups of five lines each with $|A_{\perp}\ ^{14}N| = 20 \times 10^{-4}$ cm^{-1}, attributed to a superhyperfine coupling of the unpaired copper electron to two equivalent nitrogen ligand nuclei (Senesi et al., 1985b). In conclusion, ESR analysis confirms the presence of inner-sphere complexes for Cu^{2+} in HS, which was previously suggested by means of other spectroscopic techniques and physicochemical parameters, that is, the large formation constants of Cu^{2+}–humic complexes with respect to complexes of other divalent metal ions, which are largely bound by HS as hydrated ions (Schnitzer, 1978; Stevenson, 1982).

C. Vanadyl Complexes

A representative example of a more complicated but well-resolved ESR spectrum of natural HS, obtained over a scan range of 8000 G (= 800 mT), at 77 K, for a

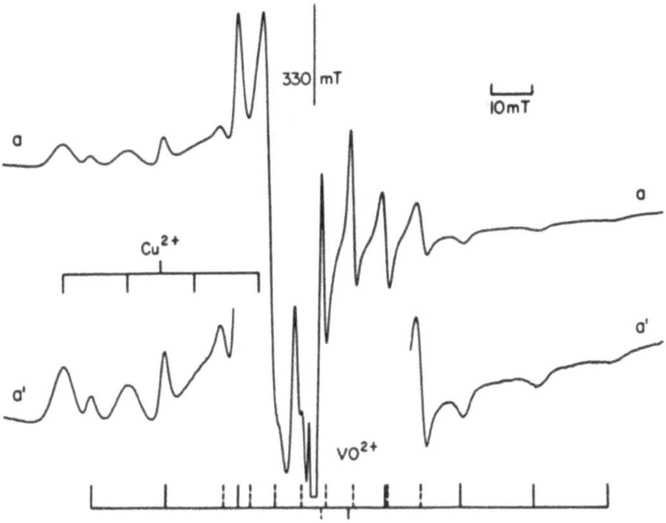

Figure 13. ESR spectrum of a loam soil humic acid powder at 77 K, scan range 2000 G; *a'* refers to a higher gain spectrum (Senesi et al., 1989b).

loam soil, powdered HA is shown in Figure 12. Analysis of the richly structured pattern at about $g = 2$, made on enlarged spectra of the region (Fig. 13), indicate the presence of two distinct, overlapping rigid-limit patterns of the "axial" type. One constitutes a major absorption at $g_\perp$ and a lesser absorption resolved in a quadruplet at $g_\parallel$, which typically indicates inner-sphere Cu^{2+}–HA complexes. The remainder of the anisotropic pattern around $g = 2$ consist of superimposition of two hyperfine octuplets, corresponding to the parallel and perpendicular components of vanadyl ions, VO^{2+} (nuclear spin of V, $I = 7/2$) held in inner-sphere complexes in the HA molecule. The ESR parameters of this spectrum ($g_\parallel = 1.944$, $|A_\parallel| = 172 \cdot 10^{-4}$ cm^{-1}, $g_\perp = 1.984$, $|A_\perp| = 68 \cdot 10^{-4}$ cm^{-1}) were consistent with 4 O donor atoms coordinating the vanadyl group in an equatorial plane (Senesi et al., 1985c, 1989c), similar to what was found for other VO^{2+}–HS complexes (McBride, 1978; Templeton and Chasteen, 1980; Abdul-Halim et al., 1981).

These results suggest that VO^{2+} complexes in soil HA are characterized by weaker ligand fields and less covalency than those in other humic materials, for example, bog-and peat-humates and soil FA fractions, whose ESR spectral parameters were consistent with VO^{2+} group in a mixed oxygen and nitrogen environment (Cheshire et al., 1977). The vanadyl ion that occurs naturally in soil HA, as well as in some soil fungal HA, therefore coordinates primarily to surface carboxylate groups while it remains partially hydrated, thus resulting in relatively labile and exchangeable forms (Senesi et al., 1985c, 1987c, 1989c).

ESR spectroscopy added to the evidence on the known ability of HS to reduce the metavanadate ion, VO^{3-}, which is the most stable form of vanadium under typical soil conditions, to VO^{2+} in acidic media and to complex the vanadyl ion in forms that are protected from oxidation at very high pH values, at which it is termodynamically unstable (Lakatos et al., 1977; Cheshire et al., 1977; Wilson and Weber, 1979; McBride, 1980a).

ESR spectroscopy has been applied to measure the stability constants and stoichiometry of VO^{2+} complexes formed by soil FA fractions separated by gel filtration chromatography (Templeton and Chasteen, 1980). The approximate molecular weights of FA fractions were assessed by combining correlation times obtained by ESR analysis and gel-filtration chromatography results. ESR data also provided information on different binding sites available to VO^{2+} ions in HS from different sources indicating that catechol groups are probably responsible for the reduction of VO_3^- to VO^{2+} by HS (Templeton and Chasteen, 1980).

D. Manganese Complexes

The ESR spectrum obtained at 77 K over a scan range of 1000 G (Fig. 14) for a powdered, decomposing chestnut leaf litter sample featured a well-resolved isotropic pattern consisting of 6 almost equally spaced principal lines and 10 secondary lines of lesser intensity. The ESR parameters (g_i = 2.005 and A_{iso} = 88 · 10^{-4} cm $^{-1}$) are consistent with outer-sphere complexes of Mn^{2+} ions (I = 5/2), which retain their inner-sphere hydration shell and are bound, principally by electrostatic forces, to six oxygen atoms of negatively charged carboxylate and phenolate functional groups of the litter in a distorted octahedral environment (Senesi and Sposito, 1989). The Mn^{2+} resonance disappeared after intense acid treatment of the litter sample. Similar Mn^{2+} patterns, although less resolved, were observed for decomposing litter samples of different origin (Senesi et al., 1989a), newly humified clover leaves (Sonbol and Baverez, 1985), HA from arable soils (Cheshire et al., 1977, 1979), and deep peat soils (Abdul-Halim et al., 1981).

ESR parameters consistent with inner-sphere complexes of Mn^{2+} ions, coordinated octahedrally with carboxyl, phenolic hydroxyl and/or carbonyl, have been measured for some peat HA (Lakatos et al., 1977). A more recent, extensive ESR investigation indicated that much of the Mn^{2+} adsorbed by soil organic matter at low pH was present in outer-sphere complexes, while a greater fraction of Mn^{2+} could enter inner-sphere as the pH and temperature were raised (McBride, 1982c).

In aqueous systems, Mn^{2+} ions can be bound simultaneously by electrostatic forces and hydrogen bonds as hexahydrated ion to a soil FA in outer-sphere, unsymmetrical sites (Gamble et al., 1977). Manganese complexed in this form would result more available to plant roots and microorganisms than when occurring in chelate forms. ESR spectroscopy has been found more sensitive, convenient and faster than the ion-exchange method in determining equilibrium constants for Mn^{2+}-FA complexes (Gamble et al., 1976).

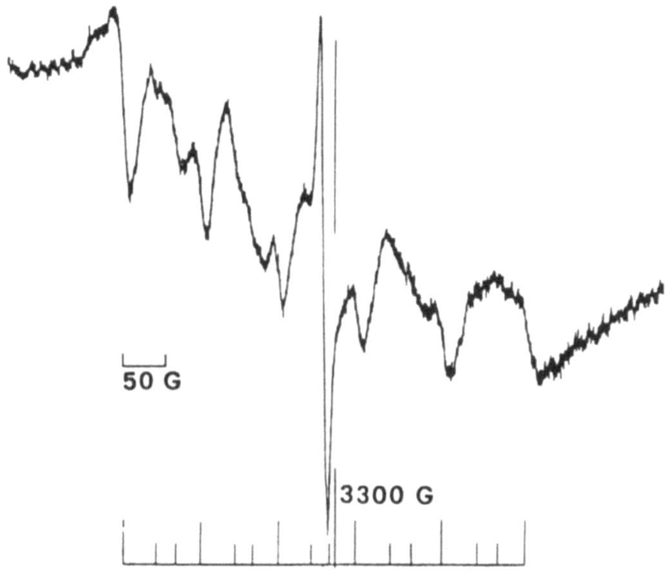

Figure 14. ESR spectrum of powdered chestnut leaf litter at 77 K, scan range 1000 G. (Reproduced with permission from N. Senesi and G. Sposito, "Characterization and stability of transition metal complexes of chestnut (*Castanea sativa* L.) leaf litter," *J. Soil Sci.* 40:461–472, 1989, by courtesy of Blackwell Scientific Publications.)

E. Molybdenum Complexes

ESR evidence has been furnished that a peat HA was able to reduce molybdate to Mo(V) and to complex Mo(V) species, in strongly acidic media (Lakatos et al., 1977; Goodman and Cheshire, 1982). The ESR spectrum of a peat HA complex with Mo(V) enriched in ^{95}Mo ($I = 5/2$) featured two components, each split into two six-line hyperfine patterns at $g_{\parallel}$ and $g_{\perp}$, consistent with two different axially symmetric Mo(V)–HA complexes (Goodman and Cheshire, 1982).

The sample obtained after 0.1 M HC1 treatment of the ^{95}Mo(V)-enriched–HA complex gave a low-intensity six-line ESR spectrum, possibly arising from Mo(III) (Goodman and Cheshire, 1982). Molybdenum (III) species can therefore form and remain stable in the solid state even in aerobic conditions when protected in HS complexes.

F. Effects of Metal Addition on Indigenous Metal Complexes

ESR spectroscopy has been usefully applied to monitoring the variations of paramagnetic metal–HA complexes that occur in a sequence of HA samples extracted from a loam soil amended with increasing rates of metal-rich sewage sludges (Senesi et al., 1989b). Representative spectra of HA isolated from the nonamended soil (Fig. 15a) and from the same soil amended with 90 Mg/ha

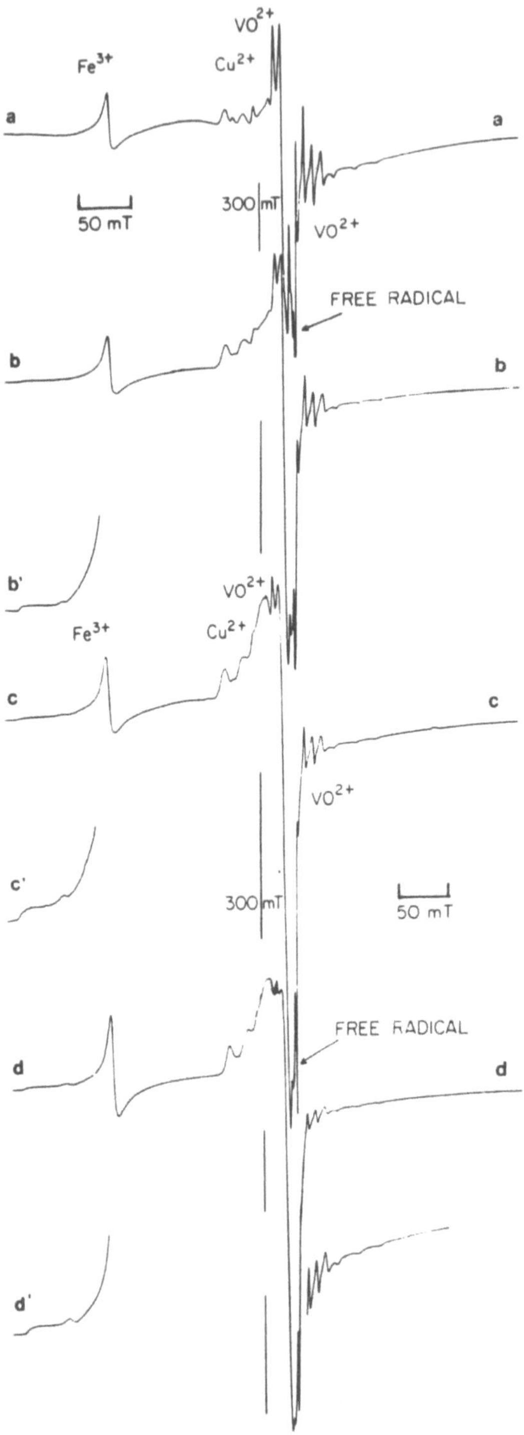

Figure 15. ESR spectra of loam soil humic acid (a) and sludge-amended soil humic acid (b, 90 Mg/ha; c, 225 Mg/ha; and d, 630 Mg/ha), at 77 K, scan range 8000 G; b′, c′, d′ refers to spectra recorded at a higher gain (Senesi et al., 1989b).

(Fig. 15b), 225 Mg/ha (Fig. 15c), and 360 Mg/ha (Fig. 15d) indicate a progressive increase, along the sequence, of the intensity of the Cu^{2+}–HA ESR pattern and, to a lesser extent, of Fe^{3+}–HA signals, while a simultaneous decrease in intensity and loss of resolution is evident for the VO^{2+}–HA pattern. This trend, confirmed by IR and total metal content analyses, suggests that increasing metal loading of soil by sludge application, results in increasing complexation of Fe^{3+} and Cu^{2+} by HA, by either partial displacement of labilly bound VO^{2+} ions or involvement of additional available sites.

Laboratory-prepared model complexes of paramagnetic metal ions and HS have been studied by ESR spectroscopy to ascertain the residual binding capacity of HS toward metal ions such as Cu^{2+}, Fe^{3+}, and Mn^{2+} used as spin probes, the chemical and geometrical structure of the complexes formed in various conditions, and their stability under physical and chemical treatments, such as proton and metal ion exchange reactions.

ESR spectra in Figure 16 furnish an example of the variations induced in the loam soil HA by interaction with solutions of Cu^{2+}, Mn^{2+}, and (Cu^{2+} + Mn^{2+} + Fe^{3+}) simultaneously, and successive exhaustive washings of the products obtained with water or 0.1 M HC1 (Senesi et al., 1990). ESR spectra of HA treated with Cu^{2+} or Mn^{2+} ions (Fig. 16, b and c) show, with respect to the spectrum of the original HA (Fig. 16 a), the disappearance of the Fe^{3+} ESR signals and the appearance of an intense and broad signal superimposed on the original hyperfine pattern at $g = 2$. This probably results from extended spin–spin interactions between uppaired electrons on neighboring Cu^{2+} or Mn^{2+} ions. Exhaustive acid treatment of these interaction products gave ESR spectra featured by the partial restoration of Fe^{3+}, VO^{2+}, and HFR resonances and either the apparent reinforcement of the Cu^{2+} pattern (Fig. 16, b_1) or the complete disappearance of the Mn^{2+} signal (Fig. 16, c_1). The ESR spectra of HA treated simultaneous with Cu^{2+} +Mn^{2+} +Fe^{3+} ions are dominated by the intense resonance line at $g = 4.2$, associated with a lesser resonance at $g = 8.5$, both typical of Fe^{3+} ions.

These results, together with data of IR and total metal analysis, indicate a high residual binding capacity of HA for the three metal ions used as probes in the order Fe > Cu > Mn. Water-stable Fe^{3+}–HA complexes also appear partially resistant to proton- and metal-exchange reactions, whereas Cu^{2+} and Mn^{2+} are desorbed almost completely in these conditions. ESR spectral parameters also indicate that, despite the variety of treatments made, the metal–HA complexes are characterized in any case by a chemical and geometric structure similar to that of the indigenous metal–HA complexes (Senesi et al., 1990).

Similar metal interaction experiments performed on HA and FA extracted from various soils and sludges or synthesized by soil fungi have provided ESR information that contributed to the understanding of HA–metal interaction mechanisms in soil and related systems (Senesi and Sposito, 1987a, 1989; Senesi et al., 1977e, 1985a,b, 1986c, 1987d, 1989a, 1990).

Recently obtained ESR evidence substantiates that the fluorescence quenching phenomena observed when Cu^{2+} ions are added to a chestnut leaf litter aqueous extract at various pH values could be ascribed to the formation of stable, discrete Cu^{2+}–litter complexes (Sposito et al., 1988). The disappearance of Fe^{3+} and Mn^{2+}

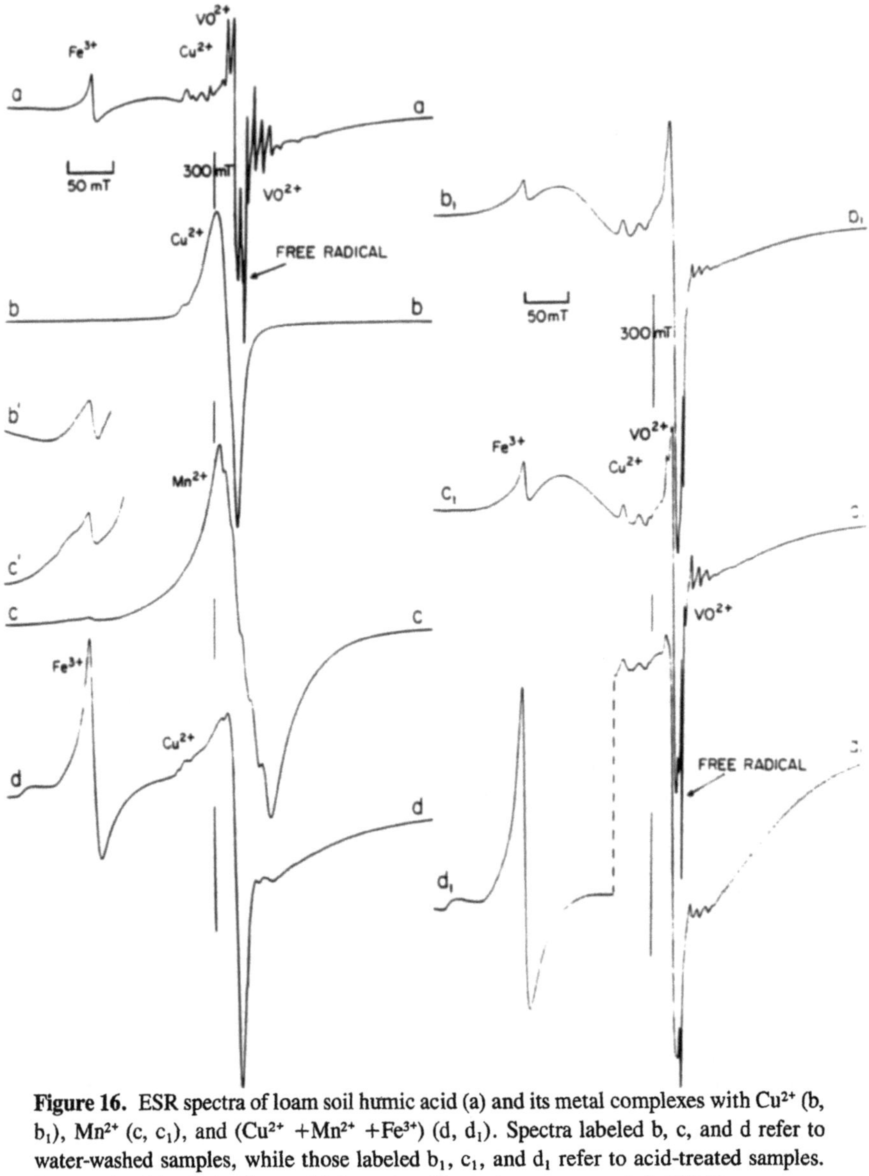

Figure 16. ESR spectra of loam soil humic acid (a) and its metal complexes with Cu^{2+} (b, b_1), Mn^{2+} (c, c_1), and (Cu^{2+} +Mn^{2+} +Fe^{3+}) (d, d_1). Spectra labeled b, c, and d refer to water-washed samples, while those labeled b_1, c_1, and d_1 refer to acid-treated samples. Superscripts b′ and c′ refer to a higher gain. All spectra were recorded at 77 K over a scan range of 8000 G. (Senesi et al., 1990).

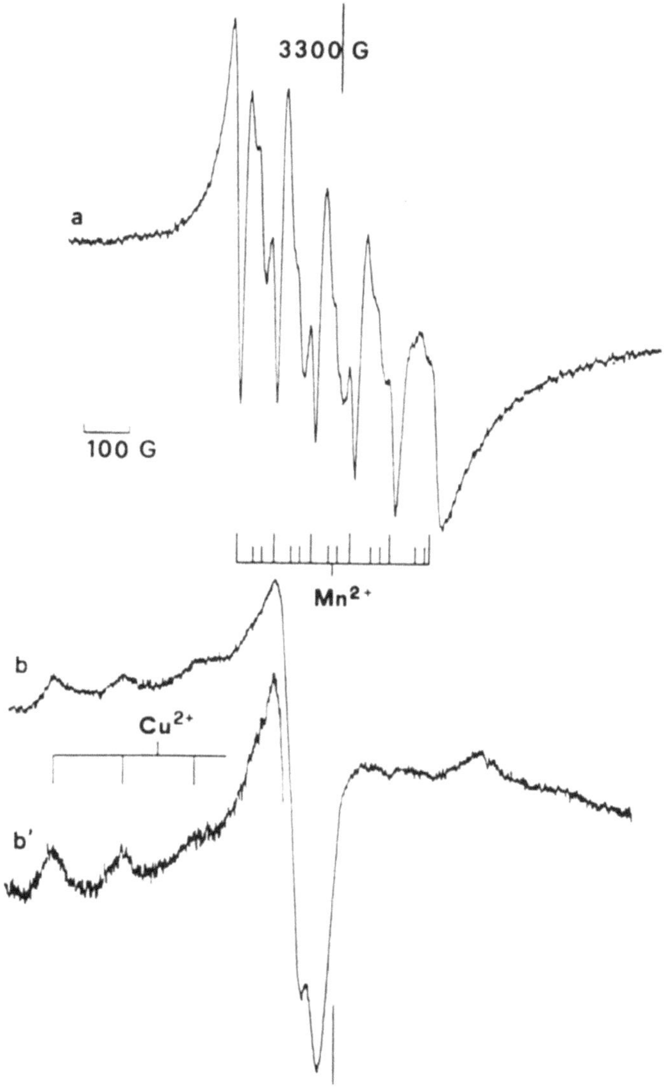

Figure 17. ESR spectra of leaf litter aqueous extract (a) and its Cu^{2+} complexes (b), recorded on frozen (77 K) solutions at 77 K over a scan range of 2000 G; a′ refers to a higher gain (Sposito et al., 1988).

signals and the appearance of the typical rigid-limit ESR pattern of Cu^{2+} ions, with the Cu^{2+}–litter interaction, confirms that Cu^{2+} ions can displace Fe^{3+} and Mn^{2+} from the litter binding sites by forming inner-sphere–type complexes (Fig. 17).

VI. Metal and/or Organic Interactions with Oxides, Hydrous Oxides, and Clay Minerals

A. Copper Spin Probes

Copper (II) is an important representative of transition metals in biological and catalytic systems and exhibits many advantageous features as a spin-probe for the application of ESR spectroscopy. ESR Cu^{2+} probes have been applied extensively to studies on adsorptive properties, surface interactions, and structural chemistry of soil and synthetic metal oxides and hydrous oxides and clay minerals. Some of the most representative information obtained on this subject is presented briefly in the following text.

ESR studies of Cu^{2+}–alumina systems revealed that noncrystalline $Al(OH)_3$ and microcrystalline AlOOH chemisorb Cu^{2+} at isolated bonding sites, probably by forming one or two direct bonds between surface Al–O groups and Cu^{2+}, which is favored at high pH (McBride, 1982a). The ability of a high surface-area gibbsite to adsorb Cu^{2+} has been studied to evaluate the influence of the planar $Al(OH)_3$ surface on the behavior of Cu^{2+} (McBride et al., 1984). The rigid-limit ESR spectrum observed after equilibration of gibbsite films in $Cu(NO_3)_2$ solutions for one day had values of $g_{\parallel}$ and $|A_{\parallel}|$ intermediate between those of $Cu(H_2O)_6^{2+}$ and $Cu(OH)_4^{2-}$, thus suggesting Cu^{2+} ions in equatorial coordination to water molecules (probably two) and hydroxyl or structural oxygen atoms. After exposure to NH_3 vapor, large decreases in $g_{\parallel}$ and $g_{\perp}$ and increase in $|A_{\parallel}|$ indicate that adsorbed Cu^{2+} may form a complex with NH_3 while remaining rigidly bound at the oxide surface. At least one Cu–O–Al bond was therefore formed in the process of chemisorption. Additional ESR results suggest the existence of two forms of Cu^{2+} oriented on gibbsite surfaces. At low pH, monomeric Cu^{2+} is adsorbed on the surface, with its symmetry axis aligned perpendicular to the gibbsite (001) faces, whereas a different form of Cu^{2+}, with a broad featureless resonance was evident above pH 4.5, becoming more predominant at higher pH as the rigid-limit spectrum diminished. At pH > 5, the gibbsite thus appeared to promote the hydrolysis and polymerization of Cu^{2+}, with further adsorption at the surface.

In a later study, McBride (1985b) investigated the effect of the chelating ligand, glycine, on the bonding of Cu^{2+} at moist surfaces of microcrystalline gibbsite and boehmite under varying pH and Cu/glycine ratio conditions. ESR results indicated that in both experiments Cu^{2+} adsorbed to form rigidly bound ternary complexes in which the metal coordinated simultaneously with a surface hydroxyl and one or more glycine molecule. The values assumed by ESR parameters were consistent with $Cu(gly)^+$, as the preferred species adsorbed on gibbsite, with the orientation of the Cu z-axis normal to the (001) surface, as shown:

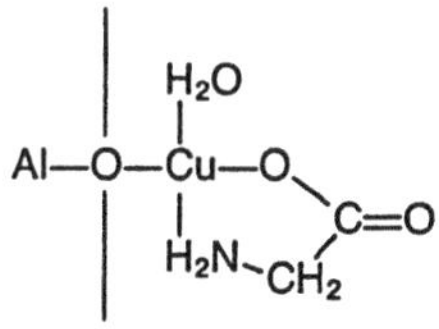

(18)

Under the same conditions, boehmite preferentially adsorbed $Cu(gly)_2^0$, with one equatorial ligand position at the Cu^{2+} ion most likely coordinated with the surface and the remaining three occupied by two amine groups and one carboxyl of glycine, while the second carboxyl occupied an axial (weak ligand) position, as shown:

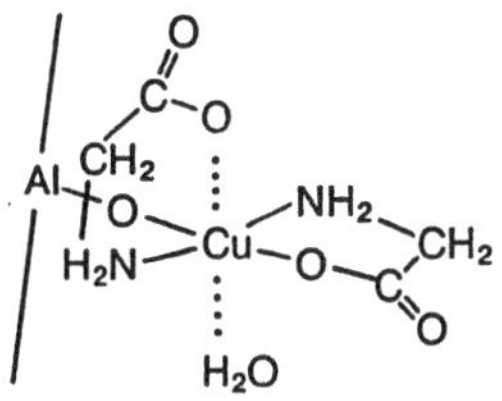

(19)

The hypothesis that one of the glycine molecules was not chelated at equatorial ligand positions of Cu^{2+} is supported by the finding that it is readily removed by water washing. A large excess of glycine in solution destabilized the ternary complex and caused the desorption of Cu^{2+}. Above pH 5, the presence of glycine reduced Cu adsorption on gibbsite, apparently by reducing Cu^{2+} hydrolysis.

By comparing ESR spectra and ESR parameters of Cu^{2+} adsorbed on a hydroxy-aluminum-hectorite complex (OH–Al–hectorite) at pH ranging from 4.5 to 7.8 and on Na– or Mg–hectorite (McBride, 1982b), Harsh et al. (1984) showed that copper adsorption on the clay complex, unlike that on hectorite, was highly pH-dependent and Cu^{2+} was chemisorbed to discrete sites, most likely on the $Al_x(OH)_y^{3x-y}$ polymers. ESR spectra of air-dried films of the Cu^{2+}–OH–Al hectorite suggested a square planar Cu^{2+} complex oriented with its z-axis perpendicular to the hectorite sheets. At pH 4.5 most of the Cu^{2+} was present as $Cu(H_2O)_4^{2+}$, while at a higher pH (7.4), the spectrum resembled that of $Cu(OH)_4^{2-}$ on alumina, suggesting deprotonation of the $Cu(H_2O)_4^{2+}$ complex or a ligand exchange mechanism, that is, complexation, with the hydroxide surface. Thus, at a pH above 7 chemisorbed Cu^{2+} remained the dominant species, even if the solubility product of $Cu(OH)_2$ was exceeded. This result contrasts with the precipitation of Cu observed by ESR on microcrystalline gibbsite above pH 5 (McBride et al., 1984) and indicates that the interlayer OH–Al on hectorite retains more Cu^{2+} on discrete sites, that is, the Al–OH phase on hectorite has a greater adsorption capacity than the gibbsite (Harsh et al., 1984).

The adsorption of three copper species, the hydrated Cu^{2+} ion, Cu^{2+}–bisglycine and Cu^{2+}–HA complex, on montmorillonite and imogolite at pH 7 was investigated by ESR spectroscopy by Goodman et al. (1984). The ESR spectra showed that both copper–glycine and copper–HA are adsorbed on the two clay minerals with the organic ligand remaining bound to the copper and that the amount and type of adsorption depend on the size and chemical nature of the ligand. The spectra of the supernatant solutions indicated that adsorption of the glycine-complexed Cu^{2+} was much less than that of uncomplexed ion and of the Cu^{2+}–HA complex for both minerals and that montmorillonite adsorbed significantly more Cu^{2+} from the HA solution than imogolite did. ESR spectral parameters showed that in any case the presence of more than one type of chemical environment for adsorbed Cu^{2+} ions, depending on their freedom of movement and their mode and site of adsorption. The qualitatively similar spectra obtained when the uncomplexed Cu^{2+} was adsorbed on either of the minerals suggests that Cu^{2+} coordinated by water and hydroxyl group are involved in adsorption. The spectra of bisglycine–Cu^{2+} complexes were characterized by two components which were accounted for adsorption at two different types of sites, that is, on interlayer and surface sites in the mineral. ESR spectrum of Cu^{2+}–HA complex with montmorillonite was similar to that of Cu^{2+}–HA, thus indicating that the copper remains associated with the HA, probably because of the tendency in montmorillonite to form aggregates bound by Cu^{2+}–HA bridges. The spectra of the Cu^{2+}–HA system with imogolite were, however, similar to those of the uncomplexed Cu^{2+} ions, thus indicating that imogolite was able to extract copper from HA. These reactions are important in soil in the vicinity of plant roots where low-molecular-weight organic acids, including amino acids, from root exudates are found in addition to humic substances (Goodman et al., 1984).

The specific adsorption of Cu^{2+} by allophane and imogolite was evaluated by Clark and McBride (1984) through a series of allophanic clay fractions from volcanic ash soils and their synthetic analogues, which allowed avoidance of interference from other hydrous metal oxides and organic matter and to probe the ligand field properties of bound Cu^{2+} using ESR spectroscopy. ESR analysis revealed that Cu^{2+} was adsorbed as a monomer on two types of surface sites, either on a preferred site consisting of adjacent AlOH groups binding Cu^{2+} by a binuclear mechanism, or on a weaker type of sites occurring at isolated AlOH or SiOH groups. These chemisorbed forms of Cu^{2+} were readily extracted by EDTA, CH_3COOH, and metals capable of specific adsorption, but were not exchangeable. In addition, NH_3 was able to readily displace H_2O and OH^- ligand of chemisorbed Cu^{2+}, leading to the formation of ternary Cu^{2+}–ammonia–surface complexes.

ESR technique has also been used by Clark and McBride (1985) and McBride (1985a) to probe the chemical environment of Cu^{2+} at the surface of natural and synthetic allophanes and noncrystalline aluminum hydroxide in the presence or absence of adsorbed phosphate, to evaluate the influence of phosphate on the bonding mechanism of Cu^{2+} on these systems. ESR results suggested that Cu^{2+} remained bonded to surface AlOH groups of allophanes whether or not phosphate was present, thus indicating that phosphate does not coordinate strongly with Cu^{2+}. A mechanism was proposed by which phosphate coordinates with the axial position of a surface-bound Cu^{2+} ion, thereby producing a ternary surface complex, as shown here (Clark and McBride, 1985):

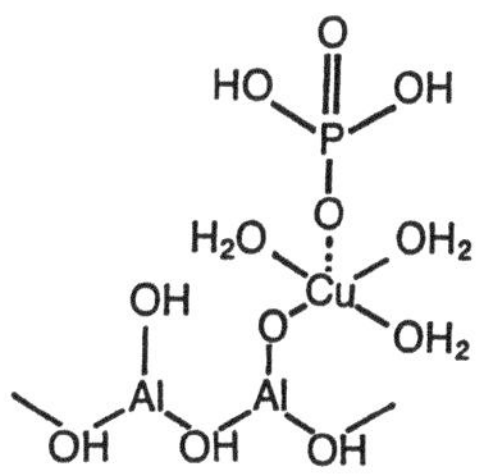

(20)

In addition, McBride (1985a) presented ESR evidence for two separate chemisorption sites for Cu^{2+} on Al hydroxide in the absence of phosphate, while high levels of sorbed phosphate suppressed Cu^{2+} adsorption, apparently by blocking the coordination of Cu^{2+} to AlOH groups on aluminum hydroxyde.

B. Anionic Organic Spin Probes

The use of anionic organic spin probes to study the adsorption process on oxides by ESR can provide information on surface-organic interaction not obtainable by other methods. The technique is sensitive, permitting adsorption of small quantities of organic anion to be studied, and allows fully wet systems to be analyzed.

The spin-probe technique has been demonstrated for the fatty acid–alumina reaction, confirming that adsorption occurs by bonding of the carboxylated end of the molecule to the surface (McBride, 1980b). The nature of organophosphate and carboxylate bonding in aqueous suspensions of noncrystalline alumina, boehmite, and gibbsite has been studied with two nitroxide spin probes with anionic functional groups (McBride, 1982d). Analysis of the ESR spectra of these molecules revealed that both were adsorbed rapidly onto the high-surface-area alumina and boehmite, whereas only the organophosphate probe was adsorbed in small amounts by the gibbsite. The very low surface area of the gibbsite sample used was possibly responsible for the lack of retention of the carboxylic acid probe (McBride, 1982d). The rotational correlation times calculated from the ESR spectra indicated that a loss in rotational motion of the molecules accompanied adsorption. The degree of motional restriction of the organic anions on the fully hydrated surfaces suggested that boehmite adsorbs the carboxylic acid largely by ligand exchange, while noncrystalline alumina adsorbs the molecule by both ligand exchange and nonspecific electrostatic processes.

C. Iron Spin Probe of Electronic Energy Storage in Clays

The capacity for electronic energy storage is considered to be one of the principal photophysical properties of clay minerals. This effect may alter mechanisms and rates of surface reactions on clay and can be evidenced by ESR as well as other methods such as thermal luminescence. To assess the extent and utility of the stored energy in surface processes of clays, it is necessary to characterize the storage sites and the luminescent centers.

Recently, Coyne and Banin (1986) examined the ESR spectra and the natural and gamma-induced thermoluminescence curves of a series of variably cation-exchanged Fe–Ca–clays prepared from Wyoming montmorillonite. The ESR signal ($g = 2$) intensity associated with the surface Fe was found to increase linearly with surface Fe content up to a nominal concentration of 50% exchangeable Fe. At >50% exchangeable Fe, no appreciable increase in the ESR signal was noted. The finding that both the thermoluminescence and the component of the ESR signal attributable to surface iron indicated a critical content of Fe suggests that an interaction probably exists between surface Fe and energy storage sites in the clay structures.

D. Thermodynamic Constants and ESR Parameters

Copper (II)–ESR probe was used to study surface complexation at the water/oxide interface and to estimate thermodynamic stability constants of the Cu^{2+}–surface complexes (Motschi, 1983; 1984). ESR parameters indicated that Cu^{2+} aquoions formed inner-sphere complexes of pseudo-square geometry with surface functional groups of hydrous-Al_2O_3, TiO_2, and some silicas.

On the basis of the well-known pronounced effect exerted on the $g_{\parallel}$-value by the variation of the functional group coordinated to Cu^{2+}, Motschi (1984) found a linear relationship between thermodynamic complex formation constants of square planar complexes of Cu^{2+} in aqueous solution and the corresponding ESR parameters $g_{\parallel}$, as a function of the ligand. The potential of the model was demonstrated in a series of ternary complexes on δ–Al_2O_3, for which speciation of Cu^{2+} and stability constants were obtained. On these bases, a major revision of current conceptions of cation adsorption in layer silicates and some catalytic systems (e.g., zeolites) was also presented (Motschi, 1984).

The model just described, capable of interrelating thermodynamics aspects with ESR spectroscopic data, is very promising for the interpretation of the chemistry of copper in soil and related organic and inorganic systems.

VII. Summary and Conclusions

The first part of this review briefly discussed the basic principles, methodology, and limitations of ESR spectroscopy. In the more extended second part, a number of examples of application of ESR technique to various natural and model systems strictly related to soil environment were reviewed. The principal type of information of interest to soil scientist provided by ESR analysis have been also described. With this respect, it should be pointed out that, although most information obtained by the ESR technique is unique, the chemical interpretation of ESR results often needs support and validation by other spectroscopic and physicochemical techniques, including IR, NMR, and fluorescence spectroscopy, electron microscopy, differential thermal analysis, and electrochemical methods.

ESR studies of organic free radicals indigenous to humic substances from soil and related systems have provided insight into some structural aspects and chemical and behavioral properties of these soil components. However, the technique has been proved more successful in probing the reactions of humic substance and mineral components of soil with their environments, that is, in elucidating the interaction and bonding mechanisms with metal ions and organic chemicals, both occurring in natural conditions and on model systems.

The ability of the ESR technique to detect transient radical species promoted in HS by various factors, such as light irradiation, pH increase, and reductive conditions, may help elucidate numerous naturally occurring processes affected by HS, such as photosensitized reactions and photodegradation of organic chemicals.

The ESR technique has been proved especially suited for the study of metal associations to HS, inorganic oxides and hydrous oxides, and clay minerals. In this respect, unique insight has been obtained on metal ion oxidation states, mechanisms of metal binding, symmetry of binding sites, identity of ligand groups, stability of metal complexes in different conditions, and mechanisms of formation and structures of ternary metal–organic–mineral associations.

Finally, results obtained by the use of ESR metal and organic probes may provide invaluable information on the role of humic and mineral soil components in the adsorption–retention–release phenomena involving metal ions and organic chemicals in the soil–water–organism system, and, in turn, on accumulation and movement, bioavailability, and biotoxicity of xenobiotic organics and metals in the soil environment.

VIII. Recommendations for Future Work

Research fields that show promise for extension and future applications of ESR technique in soil and environmental chemistry are indicated as follows. The role of free radicals needs a deeper and extensive evaluation in one-electron redox processes in natural systems, particularly as potential initiators and catalysts of the polymerization–depolymerization reaction of HS, that is, humification–mineralization processes of organic matter in soils; as antagonizers and activators of other natural organic and inorganic catalysts acting in environmental processes; as mediators in enzyme–HS associations and residual enzymatic activity in soil; as active species in photosensitization and photodegradative action of HS; in direct physiological effects exerted by HS on seed germination, root initiation, and plant growth in general; in the electron donor–acceptor properties and in the capacity of forming charge-transfer complexes of HS with organic chemicals.

Extension of ESR investigation is also needed to elucidate the mechanisms of HS-induced metal reductions in soil and aquatic systems by monitoring electrochemical redox reactions and their kinetics. More investigation is necessary into the possible correlations of ESR data and thermodynamic data of paramagnetic metal complexes, to better evaluate the metal speciation occurring in solution and the metal forms that adsorb on the solid organic and inorganic soil surfaces.

A more extended application of spin-labeling technique is also encouraged as a successful approach to estimating the dynamics of humic macromolecules in solution and adsorption–desorption surface processes on wet oxide and hydrous oxide minerals. Analysis of the line shape of the ESR spectrum of spin-labeled polymers may reveal information on the rotational correlation times of the polymer and the types of molecular motion in solution, which are related, in turn, to surface–organic interaction and to properties such as state of aggregation, molecule morphology and conformation, micelle character, and relaxation times.

The spin-trapping technique also shows great promise in examining and characterizing transient radical intermediates and relative mechanisms involved in the photochemical reactions of HS.

Spectroscopic techniques that may help overcome intrinsic difficulties of ESR spectroscopy include electron nuclear double resonance (ENDOR) and spin-echo spectroscopy. ENDOR, a combination of the ESR and NMR techniques, represents a useful tool in determining the components of hyperfine and superhyperfine matrices. ENDOR extends the resolution of the ESR experiment by resolving hyperfine contributions to inhomogeneously broadened lines in circumstances where the splitting between the hyperfine components is less than the width of the individual spin packets. Furthermore, ENDOR can be applied to studies of randomly oriented systems, such as frozen solutions, in which an ESR experiment is unable to provide the required resolution.

Electron spin-echo spectroscopy can be applied to the same situations as ENDOR, namely, where the hyperfine and superhyperfine interactions give an inhomogeneously broadened line. However, the spin-echo technique is a useful alternative to ENDOR when the splittings are small and conventional ENDOR is difficult to perform. The spin-echo experimental apparatus and spectral analysis is, however, more complex than for ENDOR.

ENDOR and spin-echo spectroscopy have not yet been applied to soil chemistry studies, and it is in this area where major scientific activity is expected to occur.

References

Abdul-Halim, A.L., J.C. Evans, C.C. Rowlands, and J.H. Thomas. 1981. An EPR spectroscopic examination of heavy metals in humic and fulvic acid soil fractions. Geochim. Cosmochim. Acta 45:481–487.

Abragam, A., and B. Bleaney. 1970. Electron Paramagnetic Resonance of Transition Ions. Clarendon Press, Oxford.

Alger, R.S. 1968. Electron Paramagnetic Resonance: Techniques and Applications. Wiley-Interscience, New York.

Atherton, N.M., P.A. Cranwell, A.J. Floyd, and R.D. Haworth. 1967. Humic acid I. ESR spectra of humic acids. Tetrahedron 23:1653–1667.

Bel'kevic, P.I., M.S. Minkevich, V.F. Stel'makh, and V.P. Strigutskii. 1974. Application of spin concentration effects to the study of the paramagnetic characteristics of humic acids by thermal degradation. Doklady Akademii Nauk. B.S.S.R. 18:432–433.

Bijl, D., H. Kainer, and A.C. Rose-Innes. 1959. Biradical molecular compounds: A study by electron spin resonance. J. Chem. Phys. 30:765–770.

Blois, M.S., Jr., H.W. Brown, and J.E. Maling. 1961. Precision g-value measurements of free radicals of biological interest. In Blois, M.S. Jr., H.W. Brown, R.M. Lemmon, R.O. Lindlom, and M. Weissbluth (eds.). Free Radicals in Biological Systems. pp. 117–131. Academic Press, New York.

Bollag, J.M. 1983. Cross coupling of humus constituents and xenobiotic substances. In Christman, R.F., and E.T. Gjessing (eds.). Aquatic and Terrestrial Humic Materials. pp. 127–141. Ann Arbor Science, Ann Arbor, MI.

Boyd, S.A., L.E. Sommers, D.W. Nelson, and D.X. West. 1981. The mechanism of copper(II) binding by humic acid. An electron spin resonance study of a copper(II)–humic acid complex and some adducts with nitrogen donors. Soil Sci. Soc. Am. J. 45:745–749.

Boyd, S.A., L.E. Sommers, D.W. Nelson, and D.W. West. 1983. Copper(II) binding by humic acid extracted from sewage sludge: An electron spin resonance study. Soil Sci. Soc. Am. J. 47:43–46.

Bresnahan, W.T., C.L. Grant, and J.H. Weber. 1978. Stability constants for the complexation of copper(II) ions with water and fulvic acids measured by an ion selective electrode. Anal. Chem. 50:1675–1679.

Bruce, J.M. 1974. Photochemistry of quinones. In Patai, S., ed., The Chemistry of Quinonoid Compounds. pp. 465–538. Wiley, New York.

Calderoni, G., and M. Schnitzer. 1984. Effects of age on the chemical structure of paleosol humic acids and fulvic acids. Geochim. Cosmochim. Acta 48:2045–2051.

Carrington, A., and A.D. McLachlan. 1967. Introduction to Magnetic Resonance with Applications to Chemistry and Chemical Physics. Harper & Row, New York.

Chen, Y., N. Senesi, and M. Schnitzer. 1977. Information provided on humic substances by E_4/E_6 ratios. Soil Sci. Soc. Am. J. 41:352–358.

Chen, Y., N. Senesi, and M. Schnitzer. 1978. Chemical and physical characteristics of humic and fulvic acids extracted from soils of the Mediterranean region. Geoderma 20:87–104.

Cheshire, M.V., M.L. Berrow, B.A. Goodman, and C.M. Mundie. 1977. Metal distribution and nature of some Cu, Mn and V complexes in humic and fulvic acid fractions of soil organic matter. Geochim. Cosmochim. Acta 41:1131–1138.

Cheshire, M.V., M.L. Berrow, B.A. Goodman, and C.M. Mundie. 1979. The nature and origin of the organic matter–metal complexes of soil. Coll. Int. C.N.R.S. n° 303, Migrations Organo-Minérales dans les Sols Tempérés, pp. 241–246.

Cheshire, M.V., and P.A. Cranwell. 1972. Electron spin resonance of humic acids from cultivated soils. J. Soil Sci. 23:424–430.

Chukov, S.N., and A.F. Gurov. 1984. Paramagnetic properties of organomineral compounds of humus illuvial sand podzol. Trudi Biol. Nauk Isledov. Ist. Leningrad Univ. 35:138–149.

Chukov, S.N., and S.I. Nikonova. 1980. Application of ESR spectroscopy to the study of organic matter in humus-eluvial-iron-podzol soils. Vestnik Leningradsk. Univ. 9:115.

Chukov, S.N., A.S. Staraev, and V.P. Tschiplenkov. 1983. Further investigations on interaction mechanisms between humic acids and Fe^{3+} ion. Vestnik Leningradsk. Univ. 15:116.

Chukov, S.N., and V.P. Tschiplenkov. 1983. Influence of purification methods of humic acid on its physico-chemical properties. Vestnik Leningradsk. Univ. 15:116.

Clark, C.J., and M.B. McBride. 1984. Chemisorption of Cu(II) and Co(II) on allophane and imogolite. Clays & Clay Minerals 32:300–310.

Clark, C.J., and M.B. McBride. 1985. Adsorption of Cu(II) by allophane as affected by phosphate. Soil Sci. 139:412–421.

Coyne, L.M., and A. Banin. 1986. Effect of adsorbed iron on thermoluminescence and electron spin resonance spectra of Ca-Fe-exchanged montmorillonite. Clays & Clay Minerals 34:645–650.

Crosby, D.G. 1976. Non-biological degradation of herbicides in soil. In Audus, L.J. (ed.). Herbicides: Physiology, Biochemistry, and Ecology, Vol. 2, 2nd ed. pp. 65–97. Academic Press, New York.

Eastman, M.P., D.E. Patterson, and K.H. Pannell. 1984. Reaction of benzene with Cu(II)–arene complexes on montmorrilonite. J. Chem. Phys. 77:784–790.

Eloranta, J., and K. Hannonen. 1985. EPR parameters as a measure of the high spin Fe(III) complex of peat. Finn. Chem. Lett. 143–145.

Eloranta, J., R. Koskinen, and P. Pohjola. 1980. EPR study on paramagnetic species appearing in peat samples from bogs in Central Finland. Finn. Chem. Lett. 161–163.

Eltantawy, I.M., and M. Baverez. 1978. Structural study of humic acids by x-ray, electron spin resonance, and infrared spectroscopy. Soil Sci. Soc. Am. J. 42:903–905.

Filip, Z., M.V. Cheshire, B.A. Goodman, and D.B. McPhail. 1985. The occurrence of copper, iron, zinc and other elements and the nature of some copper and iron complexes in humic substances from municipal refuse disposed in a landfill. Sci. Total Environ. 44:1–16.

Foster, R. 1969. Organic Charge-Transfer Complexes. Academic Press, London.

Gamble, D.S., C.H. Langford, and J.P.K. Tong. 1976. The structure and equilibria of fulvic acid and its reactions with metal ions. In P.C. Singer (ed.). Trace Metals and Metal-Organic Interactions in Natural Waters. pp. 265–302. Ann Arbor Science, Ann Arbor, Mich.

Gamble, D.S., M. Schnitzer, and D.S. Skinner. 1977. Mn(II)–fulvic acid complexing equilibrium measurements by electron spin resonance spectrometry. Can J. Soil Sci. 57:47–53.

Ghosh, K., and M. Schnitzer. 1980. Effects of pH and neutral electrolyte concentration on free radicals in humic substances. Soil Sci. Soc. Am. J. 44:975–978.

Goodman, B.A., and M.V. Cheshire. 1973. Electron paramagnetic resonance evidence that copper is complexed in humic acid by the nitrogen of porphyrin groups. Nature (New Biology) 244:158–159.

Goodman, B.A., and M.V. Cheshire. 1976. The occurrence of copper-porphyrin complexes in soil humic acids. J. Soil Sci. 27:337–347.

Goodman, B.A., and M.V. Cheshire. 1982. Reduction of molybdate by soil organic matter: EPR evidence for formation of both Mo(V) and Mo(III). Nature 299:618–620.

Goodman, B.A., H.L. Green, and D.B. McPhail. 1984. An electron paramagnetic resonance (EPR) study of the adsorption of copper complexes on montmorillonite and imogolite. Geochim. Cosmochim. Acta 48:2143–2150.

Harsh, J.B., H.E. Doner, and M.B. McBride. 1984. Chemisorption of copper on hydroxy-aluminum-hectorite: An electron spin resonance study. Clays & Clay Minerals 32:407–413.

Hayes, M.H.B. 1970. Adsorption of triazine herbicides on soil organic matter, including a short review on soil organic matter chemistry. Res. Review 32:131–174.

Hayes, M.H.B., R.S. Swift, R.E. Wardle, and J.K. Brown. 1975. Humic materials from an organic soil: A comparison of extractants and of properties. Geoderma 13:231–240.

Ingram, D.J.E. 1969. Biological and Biochemical Applications of Electron Spin Resonance. Hilger, New York.

Kallianou, Ch.S., and N.J. Yassoglou. 1985. Bonding and oxidation state of iron in humic complexes extracted from some greek soils. Geoderma 35:209–221.

Khan, S.U. 1973. Equilibrium and kinetic studies of the adsorption of 2,4-D and picloram on humic acid. Can. J. Soil Sci. 53:429–434.

Khan, S.U., and D.S. Gamble. 1983. Ultraviolet irradiation of an aqueous solution of promethryn in the presence of humic materials. J. Agric. Food Chem. 31:1090–1104.

Lagerkrantz, C., and M. Yhland. 1962. Photo-induced electron spin resonance in solutions of some electron donor-acceptor complexes. Acta Chem. Scand. 16:1043–1045.

Lagerkrantz, C., and M. Yhland. 1963. Photo-induced free radical reactions in the solution of some tars and humic acids. Acta Chem. Scand. 17:1299–1306.

Lakatos, B., T. Tibai, and J. Meisel. 1977. ESR spectra of humic acids and their metal complexes. Geoderma 19:319–338.

McBride, M.B. 1978. Transition metal binding in humic acids: An ESR study. Soil Sci. 126:200–209.

McBride, M.B. 1980a. A comparative electron spin resonance study of VO^{2+} complexation in synthetic molecules and soil organics. Soil Sci. Soc. Am. J. 44:495–499.

McBride, M.B. 1980b. Adsorption of fatty acid spin probes on amorphous alumina. J. Colloid Interface Sci. 76:393–398.

McBride, M.B. 1982a. Cu^{2+}-adsorption characteristics of aluminum hydroxyde and oxyhydroxides. Clays & Clay Minerals 30:21–28.

McBride, M.B. 1982b. Hydrolysis and dehydration reactions of exchangeable Cu^{2+} on hectorite. Clays & Clay Minerals 30:200–206.

McBride, M.B. 1982c. Electron spin resonance investigation of Mn^{2+} complexation in natural and synthetic molecules and soil organics. Soil Sci. Soc. Am. J. 46:1137–1143.

McBride, M.B. 1982d. Organic anion adsorption on aluminum hydroxides: Spin probe studies. Clays & Clay Minerals 30:438–444.

McBride, M.B. 1985a. Sorption of copper(II) on aluminum hydroxides as affected by phosphate. Soil Sci. Soc. Am. J. 49:843–846.

McBride, M.B. 1985b. Influence of glycine on Cu^{2+} adsorption by microcrystalline gibbsite and boehmite. Clays & Clay Minerals 33:397–402.

McBride, M.B. 1987. Adsorption and oxidation of phenolic compounds by iron and manganese oxides. Soil Sci. Soc. Am. J. 51:1466–1470.

McBride, M.B., A.R. Fraser, and W.J. McHardy. 1984. Cu^{2+} interaction with microcrystalline gibbsite. Evidence for oriented chemisorbed copper ions. Clays & Clay Minerals 32:12–18.

McBride, M.B., B.A. Goodman, J.D. Russell, A.R. Fraser, V.C. Farmer, and D.P.E. Dickson. 1983. Characterization of iron in alkaline EDTA and NH_4OH extracts of podzol. J. Soil Sci. 34:825–840.

Mortland, M.M., and T.J. Pinnavaia. 1971. Formation of copper(II) arene complexes on the interlamellar surfaces of montmorillonite. Nature (London) 229:75–77.

Motschi, H. 1983. Cu(II) bound to hydrous silica. EPR measurements to characterize surface coordination. Naturwiss. 70:S.519.

Motschi, H. 1984. Correlation of EPR parameters with thermodynamic stability constants for copper(II) complexes. Cu(II)–EPR as a probe for the surface complexation at the water/oxide interface. Colloids and Surface 9:333–347.

Parris, G.E. 1980. Covalent binding of aromatic amines to humates. 1. Reactions with carbonyl and quinones. Environ. Sci. Technol. 14:1099–1106.
Pinnavaia, T.J., P.L. Hall, S.S. Cady, and M.M. Mortland. 1974. Aromatic radical cation formation on the intracrystal surfaces of transition metal layer lattice silicates. J. Chem. Phys. 78:994–999.
Poole, C.P. 1967. Electron Spin Resonance. A Comprehensive Treatise on Experimental Techniques. Wiley-Interscience, New York.
Retcofsky, H.L., M.R. Hough, M.M. Maguire, and R.B. Clarkson. 1981. Nature of free radicals in coal, pyrolyzed coals, solvent refined coals and coal liquefaction products. In M.L. Gorbaty and K. Ouichi (eds.). Coal Structure, Advances in Chemistry Series 192. pp. 37–58. ACS, Washington, DC.
Riffaldi, R., and M. Schnitzer. 1972a. Electron spin resonance spectrometry of humic substances. Soil Sci. Soc. Am. J. 36:301–305.
Riffaldi, R., and M. Schnitzer. 1972b. Effects of diverse experimental conditions on ESR spectra of humic substances. Geoderma 8:1–10.
Rupert, J.P. 1973. Electron spin resonance spectra of interlamellar copper(II)–arene complexes on montmorillonite. J. Chem. Phys. 77:784–790.
Saiz Jimenez, C. 1975. Characterization del pigmento de *Eurotium echinulatum* Delacr. I. Datos analiticos. An. Edafol. Agrobiol. 34:829–839.
Saiz Jimenez, C., and F. Shafizadeh. 1984. Iron and copper binding by fungal phenolic polymers: An electron spin resonance study. Current Microbiol. 10:281–286.
Saiz Jimenez, C., and F. Shafizadeh. 1985. Electron spin resonance spectrometry of fungal melanins. Soil Sci. 139:319–325.
Schnitzer, M. 1978. Humic substances: Chemistry and reactions. In M. Schnitzer and S.U. Khan (eds.). Soil Organic Matter. pp. 1–57. Elsevier, Amsterdam.
Schnitzer, M., and K. Ghosh. 1982. Characteristics of water-soluble fulvic acid-copper and fulvic acid-iron complexes. Soil Sci. 134:354–363.
Schnitzer, M., and M. Levesque. 1979. Electron spin resonance as a guide to the degree of humification of peats. Soil Sci. 127:140–145.
Schnitzer, M., and R. Riffaldi. 1972. The determination of quinone groups in humic substances. Soil Sci. Soc. Am. J. 36:772–777.
Schnitzer, M., and S.I.M. Skinner. 1969. Free radicals in soil humic compounds. Soil Sci. 108:383–390.
Senesi, N. 1981a. Free radicals involved in electron donor-acceptor reactions between soil humic acids and photosynthesis inhibitor herbicides. Z. Pflanz. Bodenk. 144:580–586.
Senesi, N. 1981b. Spectroscopic evidence of organically-bound iron in natural and synthetic complexes with humic substances. Geochim. Cosmochim. Acta 45:269–272.
Senesi, N., D.F. Bocian, and G. Sposito. 1985a. Electron spin resonance investigation of copper(II) complexation by soil fulvic acid. Soil Sci. Soc. Am. J. 49:114–119.
Senesi, N., D.F. Bocian, and G. Sposito. 1985b. Electron spin resonance investigation of copper(II) complexation by fulvic acid extracted from sewage sludge. Soil Sci. Soc. Am. J. 49:119–126.
Senesi, N., and G. Calderoni. 1988. Structural and chemical characterization of copper, iron and manganese complexes formed by paleosol humic acids. Organic Geochem. 13:1145–1152.
Senesi, N., and Y. Chen. 1989. Interactions of toxic organic chemicals with humic substances. In Z. Gerstl et al. (eds.). Ecological Studies, Vol. 73. Toxic Organic Chemicals in Porous Media. pp. 37–90. Springer-Verlag, Berlin.

Senesi, N., Y. Chen, and M. Schnitzer. 1977a. Hyperfine splitting in electron spin resonance spectra of fulvic acid. Soil Biol. Biochem. 9:371–372.

Senesi, N., Y. Chen, and M. Schnitzer. 1977b. The role of free radicals in the oxidation and reduction of fulvic acid. Soil Biol. Biochem. 9:397–403.

Senesi, N., Y. Chen, and M. Schnitzer. 1977c. Aggregation-dispersion phenomena in humic substances in soil. Proc. Int. Symp. Soil Organic Matter Studies, Braunschweig, Sept. 6–10, 1976 Vol. II, I.A.E.A., Vienna, pp. 143–155.

Senesi, N., Y. Chen, and M. Schnitzer. 1977d. The electron and gamma-irradiation of humic substances. Fuel 56:171–176.

Senesi, N., S.M. Griffith, M. Schnitzer, and M.G. Townsend. 1977e. Binding of Fe^{3+} by humic materials. Geochim. Cosmochim. Acta 41:969–976.

Senesi, N., T.M. Miano, and J.P. Martin. 1987b. Elemental, functional and free radical characterization of humic acid-type soil fungal polymers (melanins). Biol. Fertil. Soils 5:120–125.

Senesi, N., T.M. Miano, and C. Testini. 1986a. The role of humic substances in the environmental chemistry of chlorinated phenoxyalkanoic acid and esters. in L. Pawloswki, G. Alaerts, and W.J. Lacy (eds.). Chemistry for Protection of the Environment 1985. pp. 183–196. Elsevier, Amsterdam.

Senesi, N., T.M. Miano, and C. Testini. 1987a. Incorporation of water dissolved chlorophenoxyalkanoic herbicides by humic acids of various origin and nature. In G. Giovannozzi-Sermanni and P. Nannipieri (eds.). Current Perspectives in Environmental Biogeochemistry. pp. 143–156. CNR-IPRA, Publ., Rome, Italy.

Senesi, N., G. Padovano, L. Loffredo, and C. Testini. 1986b. Interactions of amitrole, alachlor and cycloate with humic acids. Environmental Contamination, 2nd Int. Conf., Amsterdam, Sept. 1986. CEP Consultants Ltd., Publ., Edinburgh, pp. 169–171.

Senesi, N., and M. Schnitzer. 1977. Effects of pH, reaction time, chemical reduction and irradiation on ESR spectra of fulvic acid. Soil Sci. 123:224–234.

Senesi, N., and M. Schnitzer. 1978. Free radicals in humic substances. In Krumbein, E.W. (ed.). Environmental Biogeochemistry and Geomicrobiology. pp. 467–481. Ann Arbor Science, Ann Arbor, Mich.

Senesi, N., and G. Sposito. 1984. Residual copper(II) complexes in purified soil and sewage sludge fulvic acids: An electron spin resonance study. Soil Sci. Soc. Am. J. 48: 1247–1253.

Senesi, N., and G. Sposito. 1987a. Manganese (II) complexation by humic acids from soils and soil fungi. In S.E. Lindberg and T.C. Hutchinson (eds.). Heavy Metals in the Environment, Vol. 2. pp. 330–333. CEP Consultants Ltd, Publ., Edinburgh.

Senesi, N., and G. Sposito. 1987b. Copper(II) complexes with refractory anionic surfactants found in sewage sludge. Water, Air, Soil Pollut. 35:241–252.

Senesi, N., and G. Sposito. 1989. Characterization and stability of transition metal complexes of chestnut (*Castanea sativa* L.) leaf litter. J. Soil Sci. 40:461–472.

Senesi, N., G. Sposito, and G.R. Bradford. 1989a. Iron, copper and manganese complexation by forest leaf litter. Forest Sci. 35:1040–1057.

Senesi, N., G. Sposito, G.R. Bradford, and K.M. Holtzclaw. 1990. Iron, copper, and manganese complexation by humic acid extracted from an aridisol amended with sewage sludge. Water, Air, Soil Pollut. (submitted).

Senesi, N., G. Sposito, K.M. Holtzclaw, and G.R. Bradford. 1989b. Chemical properties of metal-humic acid fractions of a sewage sludge-amended aridisol. J. Environ. Qual. 18:186–194.

Senesi, N., G. Sposito, and J.P. Martin. 1985c. Complexation of some transition metal ions in soil humic acids: An ESR study. In T.D. Lekkas (ed.). Heavy Metals in the Environment, Proc. Int. Conf., Athens, Sept. 10–13, 1985. pp. 478–480. CEP Consultants Ltd., Publ., Edinburgh.

Senesi, N., G. Sposito, and J.P. Martin. 1986c. Copper(II) and iron(III) complexation by soil humic acids: An IR and ESR study. Sci. Total Environ. 55:351–362.

Senesi, N., G. Sposito, and J.P. Martin. 1987c. Naturally-occurring copper, iron and vanadyl complexes in humic acid-type polymers (melanins) from soil funti: An ESR study. In G. Giovannozzi-Sermanni and P. Nannipieri (eds.). Current Perspectives in Environmental Biogeochemistry. pp. 295–308. CNR-IPRA, Publ., Rome, Italy.

Senesi, N., G. Sposito, and J.P. Martin. 1987d. Copper(II) and iron(III) complexation by humic acid-like polymers (melanins) from soil fungi. Sci. Total Environ. 62: 241–252.

Senesi, N., and C. Steelink. 1989. Application of ESR spectroscopy to the study of humic substances. In M.H.B. Hayes, P. MacCarthy, R.L. Malcolm, and R.S. Swift (eds.). Humic Substances II. In Search of Structure. pp. 373–407. Wiley, Chichester, UK.

Senesi, N., and C. Steelink. 1990. Application of ESR spectroscopy to the study of humic substance interactions with organic xenobitics and metal ions. In P. MacCarthy, M.B.H. Hayes, R.L. Malcolm, and R.S. Swift (eds.) Humic Substances III. Interactions with Metals, Minerals, and Organic Chemicals. Wiley, Chichester, UK (in press).

Senesi, N., and C. Testini. 1982. Physico-chemical investigations of interaction mechanisms between s-triazine herbicides and soil humic acids. Geoderma 28:129–146.

Senesi, N., and C. Testini. 1983a. The environmental fate of herbicides: The role of humic substances. In R. Hallberg (ed.). Environmental Biogeochemistry, Proc. 5th Int. Symp., Stockholm, June 1–5, 1981. Ecological Bull. 35, pp. 477–490.

Senesi, N., and C. Testini. 1983b. Spectroscopic investigation of electron donor-acceptor processes involving organic free radicals in the adsorption of substituted urea herbicides by humic acids. Pesticide Sci. 14:78–89.

Senesi, N., and C. Testini. 1984. Theoretical aspects and experimental evidence of the capacity of humic substances to bind herbicides by charge-transfer mechanisms (electron donor-acceptor processes). Chemosphere 13:461–468.

Senesi, N., C. Testini, and D. Metta. 1984. Binding of chlorophenoxyalkanoic herbicides from aqueous solution by soil humic acids. Proc. Int. Conf. Environmental Contamination, London, July 10–13, 1984. CEP Consultants Ltd, Publ., Edinburgh. pp. 96–101.

Senesi, N., C. Testini, and T.M. Miano. 1987e. Interaction mechanisms between humic acids of different origin and nature and electron donor herbicides: A comparative IR and ESR study. Organic Geochem. 11:25–30.

Senesi, N., C. Testini, and M. Polemio. 1983. Chemical and spectroscopic characterization of soil organic matter fractions isolated by sequential extraction procedure. J. Soil Sci. 34:801–813.

Slawinska, D., and J. Slawinski. 1975a. Spectroscopic study on mild oxidation of humus acids. I. Oxidation with molecular oxygen. Polish J. Soil Sci. 8:34–47.

Slawinska, D., and J. Slawinski. 1977b. Spectroscopic study on mild oxidation of humus acids. II. Influence of light on oxidation with molecular oxygen. Polish J. Soil Sci. 8:49–58.

Slawinska, D., J. Slawinski, and T. Sarna. 1975. The effect of light on the ESR spectra of humic acids. J. Soil Sci. 26:93–99.

Slawinski, J., W. Puzyna, and D. Slawinska. 1978. Chemiluminescence in the photooxidation of humic acids. Photochem. Photobiol. 28:75–81.

Sonbol, H.A., and M. Baverez. 1985. Electron paramagnetic resonance of manganese(II) in newly humified clover leaves. Soil Sci. Soc. Am. J. 49:1169–1172.

Sposito, G., N. Senesi, and K.M. Holtzclaw. 1988. Fluorescence quenching and copper complexation by a chestnut leaf litter extract: Spectroscopic evidence. Soil Sci. Soc. Am. J. 52:632–636.

Steelink, C. 1964. Free radical studies of lignin degradation products and soil humic acids. Geochim. Cosmochim. Acta 28:1615–1622.

Steelink, C. 1966. Electron paramagnetic resonance studies of humic acid and related model compounds. Coal Science 7:80–89.

Steelink, C. 1987. Review of ESR spectroscopy of humic substances. In R.L. Wershaw and M.A. Mikita (eds.). NMR of Humic Substances and Coal. Techniques, Problems and Solutions. pp. 47–72. Lewis Publ. Inc., Chelsea, Mich.

Steelink, C., and G. Tollin. 1962. Stable free radicals in soil humic acid. Biochim. Biophys. Acta 59:25–34.

Steelink, C., and G. Tollin. 1966. Biological polymers related to catechol: Electron paramagnetic resonance and infrared studies of melanin, tannin, lignin, humic acid and hydroquinones. Biochim. Biophys. Acta 112:377–379.

Steelink, C., and G. Tollin. 1967. Free radicals in soil. In A.D. McLaren and G.M. Peterson (eds.). Soil Biochemistry. pp. 147–169. Dekker, New York.

Stevenson, F.J. 1972. Organic matter reactions involving herbicides in soil. J. Environ. Qual. 1:333–343.

Stevenson, F.J. 1982. Humus Chemistry: Genesis, Composition, Reactions. Wiley, New York.

Stott, D.E., J.P. Martin, D.D. Focht, and H. Haider. 1983. Biodegradation, stabilization in humus, and incorporation into soil biomass of 2,4-D and chlorocatechol carbons. Soil Sci. Soc. Am. J. 47:66–70.

Stuermer, D.H., K.E. Peters, and I.R. Kaplan. 1978. Source indicators of humic substances and proto-kerogen. Stable isotope ratios, elemental compositions and electron spin resonance spectra. Geochim. Cosmochim. Acta 42:989–997.

Suflita, J.M., M.J. Loll, W.C. Snipes, and J.M. Bollag. 1981. Electron spin resonance study of free radicals generated by a soil extract. Soil Sci. 131:145–150.

Swartz, H.M., J.R. Bolton, and D.C. Borg. 1972. Biological Applications of Electron Spin Resonance. Interscience, New York.

Templeton, G.D., and N.D. Chasteen. 1980. Vanadium-fulvic acid chemistry: Conformational and binding studies by electron spin probe techniques. Geochim. Cosmochim. Acta 44:741–752.

Tichy, V. 1971. Biological activity of UV-irradiated humic acids. Trans. Int. Symp. Humus et Planta V, Prague, Sept. 13–17, 1971. pp. 553–556.

Tollin, G., T. Reid, and C. Steelink. 1963. Structure of humic acid. IV. Electron paramagnetic resonance studies. Biochim. Biophys. Acta 66:444–447.

Varadachari, C., A.K. Borman, and K. Ghosh. 1983. Electron spin resonance and scanning microscopy of humic substances. J. Indian Soc. Soil Sci. 31:28–30.

Wang, T.S.C., M.C. Wang, Y.L. Ferug, and P.M. Huang. 1983. Catalytic synthesis of humic substances by natural clays, silts and soils. Soil Sci. 13:350–360.

Weed, S.B., and J.B. Weber. 1974. Pesticide-organic matter interactions. In W.D. Guenzi (ed.). Pesticides in Soil and Water. pp. 39–66. Soil Sci. Soc. Am., Madison, Wisc.

Wertz, J.E., and J.R. Bolton. 1972. Electron Spin Resonance: Elementary Theory and Practical Applications. McGraw-Hill, New York.
Wikander, G., and B. Nordén. 1988. Investigations of paramagnetic species in peat. Soil Sci. 145:289–297.
Wilson, S.A., and J.H. Weber. 1977. Electron spin resonance analysis of semiquinone free radicals of aquatic and soil fulvic and humic acids. Anal. Lett. 10:75–84.
Wilson, S.A., and J.H. Weber. 1979. An EPR study of the reduction of vanadium(V) to vanadium(IV) by fulvic acid. Chem. Geol. 26:345–354.

Microelectrophoresis Applied to the Surface Chemistry of Clay Minerals[1]

J.B. Harsh and Shihe Xu

[1]Department of Agronomy and Soils, College of Agriculture and Home Economics Research Center Washington State University, Pullman, WA 99164-6420. Project 0385.

Advances in Soil Science, Volume 14

I. Introduction

Microelectrophoresis has been used for more than 60 years to study the behavior of aluminosilicate clays, yet there is still widespread disagreement concerning the ability of this simple measurement to describe the nature of the aqueous/solid interface. The disagreement stems from questions regarding the validity of the equations used to transform the measured property, electrophoretic mobility (μ), into the zeta potential (ζ). In electrophoresis, ζ is the average electrostatic potential at the shear plane between a hydrated particle moving in response to an electric field and the stationary water through which the particle moves. To make the mathematical analysis of the forces involved in electrophoresis tractable, the shear plane must be defined as an imaginary plane separating the hydration sheath of the moving particle from the bulk water in which it is moving (Hunter, 1981).

For brevity, this review is limited in certain respects. First, we emphasize the most recent experimental data, including work in progress in our laboratory, because much of this has not been generally available. Second, most of the discussion of quantitative applications focuses on the 2:1 clay minerals because they deviate considerably from the behavior of the more fully characterized oxides and other constant potential surfaces. Finally, no attempt is made to carry out laborious derivations of the principal equations of the double layer and electrophoresis as this has been done thoroughly elsewhere (e.g., Hunter, 1981; Low, 1987; Sposito, 1984).

The first part of this chapter examines the various quantitative approaches to the problem of calculating ζ and the surface charge density at the shear plane (σ_ζ). The application of these approaches to the prediction of ion distribution and swelling forces in the interfacial region is presented. In the second part, we demonstrate the application of microelectrophoresis to increase our understanding of qualitative phenomena when model assumptions of potential distribution are not required. The specific examples of such phenomena to be discussed include demixing on smectites, ion selectivity on clay minerals, and structural models of allophane and imogolite.

II. Quantitative Applications

A. Theoretical Considerations

1. The Stern Model

A simple representation of the Stern model of the solid–solution interface (Stern, 1924; Shainberg and Kemper, 1966) is presented in Figure 1. According to this model, the surface is separated from the bulk solution by an interfacial region in which the electrostatic potential varies as a function of distance from a charged surface and depends on the surface charge and ion composition of the solution. The interfacial region is further divided into two regions—the diffuse and Stern layers—depending on the nature of ion–surface interactions. In the diffuse layer

Figure 1. Potential and charge distribution at the aqueous–solid interface according to the Stern model.

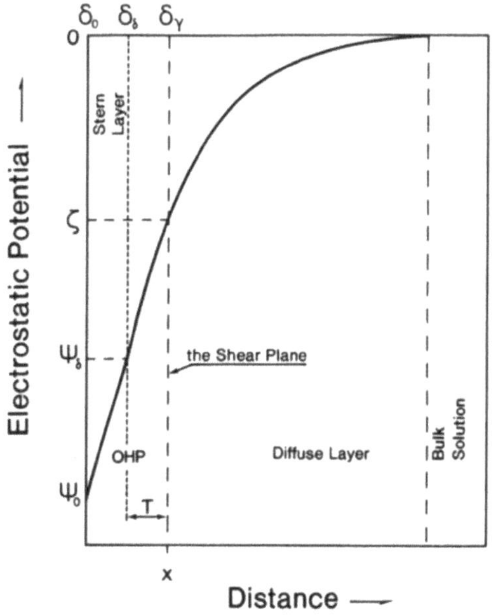

the interaction between an ion and a charged surface can be described by purely electrostatic forces (i.e., the ion may be approximated by a point charge or, at best, a hard sphere), whereas in the Stern layer, "chemical forces" such as ion–dipole interactions or covalent bonding must be invoked. The latter interactions are grouped under the term "surface complexation" and are assumed to be restricted to a region relatively close to the surface (~ 0.5 nm). The plane separating the Stern and diffuse layers is referred to as the Outer Helmholtz Plane (OHP), and both regions are defined by a higher concentration of counterions than that found in the bulk solution. The other plane represented in Figure 1 is the shear plane between the moving hydrated particle and the stationary bulk solution. The shear plane may or may not be coincident with the OHP for a given system.

The electric field that surrounds a charged clay particle suspended in an aqueous medium will govern electrostatic interactions between adjacent clay particles and between the clay surface and ions dissolved in the water. When the distance from the clay surface exceeds the diameter of a hydrated ion (~ 1.0 nm), many agree that the Poisson–Boltzmann equation may be used to describe the rate of change of the electric potential gradient (Sposito, 1984; Hunter, 1981). The region in which the Poisson–Boltzmann equation applies is referred to as the diffuse layer and the equation is as follows:

$$\frac{d^2 \psi}{dx^2} = - \frac{1}{\varepsilon_0 D} \sum_i c_i F Z_i \exp(-Z_i F \psi(x)/RT) \qquad (1)$$

where ψ is the average electrostatic potential in volts at a distance x measured normal to the mineral surface plane, ε_0 is the permittivity of a vacuum in coulombs (C) $V^{-1}\ m^{-1}$, and D is the dielectric constant of liquid water ($\varepsilon/\varepsilon_0 = D$, when ε is the permittivity of liquid water. The sum is over the concentration (c_i, in mol m^{-3}) and charge (Z_i) of all ions (i) in the solution, F is Faraday's constant, R is the molar gas constant, and T is the absolute temperature. It would be of great value to evaluate $\psi(x)$ at any point in a clay suspension. For example, we could calculate the electrostatic contribution to the swelling pressure between two clay particles, or the coulombic attraction of a cation at a given distance from a negatively charged smectite; however, no direct measurement will give us this value. As a result we must depend on indirect measurements of quantities from which $\psi(x)$ can be estimated with an appropriate model.

Electrophoretic mobility (μ) is one quantity that allows such an estimation if several simplifying assumptions are made. For example, it must be assumed that the OHP and shear plane are coincident, that the Poisson–Boltzman equation is valid to describe the potential distribution from the surface of shear to the bulk solution, and that the viscosity and dielectric constant of the fluid are either constant throughout the diffuse layer or successfully described by an appropriate model. These and other assumptions regarding the particle geometry and chemistry are discussed in the following paragraphs.

2. Calculation of ζ from μ

The first problem that arises from the calculation of ζ is assessing the validity of the assumptions made to transform the physical measurement of μ into a potential. The relationship between ζ and μ is given by Henry's (1931) equation (reported by Hunter, 1981):

$$\mu = \frac{2\varepsilon_0 D\zeta}{3\eta} f(\kappa, a) \qquad (2)$$

The electrophoretic mobility (μ) is defined as U/E, where U is the measured velocity of the particle and E is the applied electric field strength, η is the viscosity and permittivity of the medium, and $f(\kappa, a)$ is a correction factor for the fractional force and electrophoretic retardation. Electrophoretic retardation comes about because the counterions and water in the ion atmosphere surrounding the moving particle are pulled in the opposite direction, slowing the particle velocity. For small particles, the retardation can be ignored, whereas for large particles, the frictional force can be neglected. The correction factor, therefore, is a function of both the particle geometry (a = radius of curvature) and the "thickness" of the double layer ($1/\kappa$) surrounding the particle (see, e.g., Hunter, 1981, p. 27). The Debye-Hückel parameter (κ) depends primarily on the ion concentration in the bulk solution and is defined as

$$\kappa = \frac{e^2 \Sigma c_i^0 Z_i^0}{\varepsilon_0 DRT} \qquad (3)$$

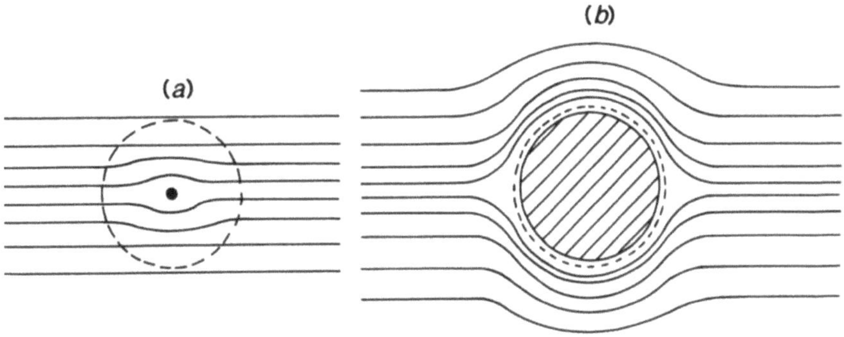

Figure 2. Effect of a nonconducting particle on the applied field. (a) $\kappa a \ll 1$; (b) $\kappa a \gg 1$. The broken line is at a distance $1/\kappa$ from the particle surface. The solid lines represent the lines of force from the applied field. (From Hunter, 1981; reproduced with permission.)

where e is the electron charge, c_i^0 is the concentration of ion i in the bulk solution, Z_i is its charge, and the other terms are as defined for Equation 1. When $\kappa a \ll 1$, the particle radius is small relative to the thickness of the double layer, deformation of the applied field is negligible (Fig. 2a), and $f(\kappa,a) = 1$. In this case, Henry's equation reduces to the Hückel equation. At the other extreme, $\kappa a \gg 1$, the field is parallel to the particle at all points along the particle surface (Hunter, 1981, p. 69, or Sposito, 1984, p. 98; Fig. 2b), $f(\kappa,a) = 1.5$, and the Smoluchowski equation results

$$\zeta = \frac{\varepsilon_0 D\mu}{\eta}$$

It is this last equation that is most familiar to soil scientists because the simplifying assumptions are often made that soil clay minerals are perfectly flat ($a = \infty$), that they do not rotate as they move in the electric field, and that relaxation effects, caused by the distortion of the field about the moving particle, are negligible. It is usually also assumed that D and η of water are constant in the surface region and equal to their values in the bulk.

Recently, several authors have come to question these assumptions based on the fact that smectite particles are actually disc-shaped particles of finite size to which the Smoluchowski equation probably does not apply (Delgado et al., 1985; Pashley, 1985; Sposito, 1987; Xu, 1988). Because there is currently no equation that suitably relates μ of a disc to its ζ potential, these authors have chosen to approximate the particle geometry with an effective spherical radius. The relationship between μ and ζ for spherical particles for which $\kappa a \simeq 1$ is complicated by the fact that $f(\kappa,a) \neq 1$ or 1.5 and neither frictional nor retardation effects can be neglected.

O'Brien and White (1978) developed a computer program that calculates ζ from μ for any rigid, nonconducting sphere in an aqueous KCl solution by a numerical method. Although other solutions to the problem of calculating ζ for

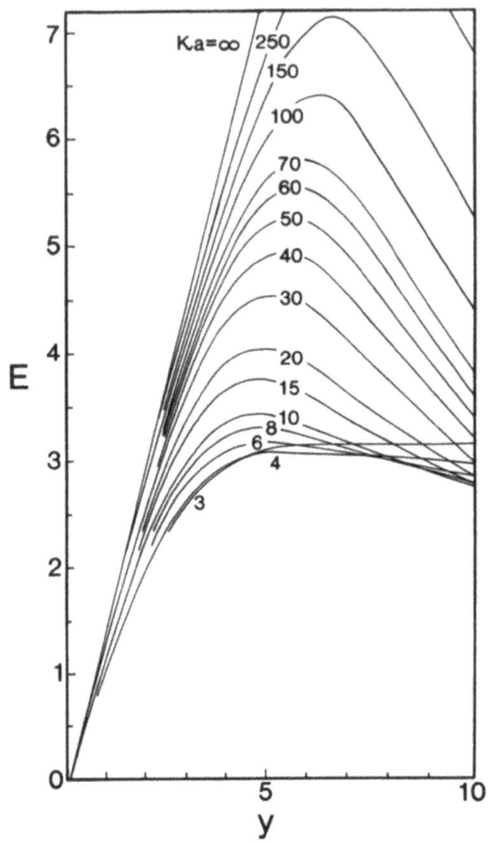

Figure 3. Variation of dimensionless mobility (E) with the reduced ζ potential ($y = e\zeta/kt$) in KCl solutions of $\kappa a > 3$. (From O'Brien and White, 1978; reproduced with permission.)

a sphere have been developed (Wiersema et al., 1966; Ottewill and Shaw, 1972), O'Brien and White's approach is the first effective means of calculation for all spherical particles regardless of temperature, particle radius, or ion concentration. This approach also accounts for the relaxation effect, which results from the continuous reformation of the ion atmosphere around the moving particle (Hunter, 1981). Relaxation occurs as the counterions reposition themselves in the diffuse layer and, thus, depends on the mobility and charge of the counterions. Ohshima et al. (1983) have recently presented an analytic expression that closely approximates the complete O'Brien and White approach but allows the calculation to be made for any electrolytes whose conductivity values are available.

A simple example will serve to show how the application of O'Brien and White's (1978) approach to a clay mineral may affect the calculation of ζ if its effective spherical diameter is used in place of the $a = \infty$ assumption. Consider a smectite particle as an ellipsoidal disc with a thickness of 3 nm and a disc radius of 300 nm. Then, $a \approx [1/3(2 \times 300^2 + 3^2)]^{1/2} = 245$ nm. For a 1:1 salt concentration of 0.1 mol m^{-1}, $1/\kappa = 30$ nm and $\kappa a = 8$. Reference to Figure 3 shows that significant error may result from assuming $\kappa a = \infty$ for a disc-shaped smectite particle.

One interesting feature of the O'Brien and White (1978) calculation is the existence of maxima in the magnitude of μ when $\kappa a > 3$ (Fig. 3). This maximum occurs at a critical ζ where the distortion of the ion atmosphere about the particle leads to a decrease in $|\mu|$ with increasing $|\zeta|$ (Midmore and Hunter, 1988). This maximum is relevant to the calculation of ζ for clay minerals when $5 \times 10^{-4} < c_i^0 < 10^{-2}$ M. Thus, two values of ζ which are proper solutions for a given μ always exist, and the choice between them is arbitrary. The availability of O'Brien and White's program, with its alternative to the confines of the Smoluchowski equation, has no doubt contributed to the experimentation by clay scientists. In the following sections we look at the results produced by the Smoluchowski equation and other approaches when applied to the calculation of several surface properties of clays. It must be kept in mind that neither approach represents an exact solution, because the geometry of a smectite particle in solution cannot be truly represented by either a sphere or an infinite, perfectly flat plate. Furthermore, it seems likely that a moving smectite sheet would tumble or rotate and, therefore, should be treated similarly to a cylinder whose orientation to the field must be averaged (Stigter, 1978).

3. Calculation of σ_ζ from ζ

Before looking at the applications of microelectrophoresis to clay minerals, we need to consider the equation relating surface charge to the electrostatic potential at that surface. For the charge at the shear plane of a particle (σ_ζ) the following equation applies (Sposito, 1984, p. 98):

$$\sigma_\zeta = \mathrm{sgn}(\zeta)\{2\varepsilon_0 DRT\sum_i c_i[\exp(-Z_i F\zeta/\mathrm{RT}) - 1]\}^{\frac{1}{2}} \qquad (5)$$

where sgn(ζ) is positive when $\zeta > 0$ and negative when $\zeta < 0$. Unlike Equation 2, which only requires that ζ be a solution of the Poisson equation, the calculation of σ_ζ requires that ζ be a solution to the Poisson–Boltzmann equation (Sposito, 1984, p. 98). This means that the validity of Equation 2 depends on the additional assumptions that (1) the shear plane behaves as an infinite, flat plane with uniform charge distribution, (2) the ions can be considered as point charges without finite size, (3) the surrounding water has constant permittivity equal to its bulk value, and (4) the ions are only acted on by long-range coulombic forces.

These assumptions are, in all probability, not valid at the solid–aqueous interface [$x = 0$ in Eq. 1] of most clay minerals because the electrical potential predicted by the Poisson–Boltzmann equation at this interface is generally high enough that inner-sphere and outer-sphere complexes are likely to form, invalidating assumption 4 (Sposito, 1984, p. 160). Additional cooperative forces among the solvent, ions, and surface will contribute to the overall picture very near the solid surface. As long as the shear plane occurs at or beyond the OHP of the particle, however, Equation 5 should be generally valid. In other words, the four assumptions stated previously are more likely to describe the conditions of ion–surface interactions at the shear plane and beyond than near the solid clay surface. Hunter (1981, p. 45) claims that corrections for ion size, permittivity, ionic polarization, and other effects result in less than a 2% correction to the simple Gouy-Chapman theory when the diffuse layer is assumed to begin at or very near the shear plane.

B. Microelectrophoresis Applied to the Determination of $\psi(x)$

1. Effect of Background Electrolyte Concentration on μ

Several studies have shown the effect of bulk electrolyte concentration on μ, but the results are inconsistent. Mattson (1929ab) was the first to perform microelectrophoresis experiments on clay minerals and used the montmorillonite clay fraction of a Sharkey soil saturated with various monovalent cations. For the Na–montmorillonite, Mattson found that μ was independent of electrolyte concentration, but he used an electrodialysis technique to first prepare H–montmorillonite from which the Na–montmorillonite was prepared. The rapid decomposition of H–montmorillonite results in a heteroionic clay consisting of Al^{3+}, Mg^{2+}, and other polyvalent cations from the clay lattice (Barshad and Foscolos, 1970). Subsequent hydrolysis of the aluminum results in hydroxy-aluminum polymers that are irreversibly adsorbed to the clay and can be expected to affect its surface properties. More recent studies have attempted to avoid or eliminate this problem; however, it is unlikely that it can be eliminated altogether.

Lyons et al. (1981) aged wet-ground mica flakes in pH 3, 7, and 10 solutions. Even the pH 7 samples showed significant increases in μ with 7 days of aging, including a charge reversal from negative to positive μ as the pH dropped from 6 to 5. They also observed that μ was pH-dependent even when determined immediately after immersion in the pH 7 electrolyte. Lyons et al. (1981) also determined ζ in a streaming potential apparatus[1] and noted that the ground mica flakes behaved like constant potential surfaces, whereas untreated basal surfaces of freshly cleaved mica behaved like constant charge surfaces. They concluded that the surfaces of the flakes were dominated by hydrolyzed Al released from the lattice during grinding and aging.

These effects should be less pronounced for smectites because (1) decomposition is much less rapid for clays of lower charge (Barshad and Foscolos, 1970) and (2) the high specific surface area of smectites compared to ground micas will decrease the relative amount of contaminant (e.g., Al) on the surface. Nevertheless, differences between various investigators' results of μ still exist and may be due, in part, to the decomposition (hydrolysis) of the homoionic clay. For an illustration of such differences, we look at the dependence of μ on electrolyte concentration.

[1]The streaming potential determination of ζ differs from microelectrophoresis in that the surface is immobile relative to the fluid (e.g., water is passed through a capillary tube of the desired surface).

Table 1. Electrophoretic mobility (μ) of Na–montmorillonite at pH 6.0 as a function of NaCl concentration

	[(μm s^{-1}) (V cm^{-1})]		
[NaCl] *M*	Callaghan and Ottewill (1974)[a]	Miller (1984)[b]	Xu (1988)[a]
2.5×10^{-6}	–	–	−3.6±0.2
3.5×10^{-5}	–	–	−3.0±0.2
10^{-4}	−3.1±0.2	−4.7	−2.7±0.1
5.0×10^{-4}	–	−4.6	−2.1±0.1
10^{-3}	−2.8±0.2	−4.7	−2.1±0.1
5.0×10^{-3}	–	−4.8	−2.5±0.2
10^{-2}	−2.7±0.2	−4.8	−2.7±0.2
10^{-1}	−3.5±0.2	–	–

[a] Clay spur montmorillonite.
[b] Upton montmorillonite.

Three studies of initially homoionic M-smectites (M = a monovalent cation, usually alkali) in which μ was determined at various electrolyte concentration are presented in Table 1. The average values of μ may differ as a result of differences among the clays themselves or the measuring apparatus, but the presence or absence of a dependence on electrolyte concentration is not as easy to explain. Miller (1984) suggests that such differences arise from the preparation procedure, and this is confirmed by Lyon's work on pH-dependence cited earlier.

The major concern in preparing homoionic clays is the release of Al, Mg, and other octahedral cations during the saturating and washing steps. Calculations based on the solubility of montmorillonite and reasonable selectivity coefficients for Na → Al exchange indicate that Al alone will occupy more than 10% of the exchange sites (J.A. Kittrick, unpublished data). The Al may subsequently hydrolyze and become irreversibly adsorbed to the cation exchange sites. Two of the factors that contribute to this release are (1) low or high pH (Lyons et al., 1981) and (2) low bulk solution concentration (Bar-On and Shainberg, 1970). Of the three studies presented in Table 1, only Xu's (1988) preparation was likely to produce a montmorillonite dominated by exchangeable Na, at least, at [Na] > 10^{-3} *M*. Both other procedures involved centrifugation or dialysis in distilled or deionized water, which results in replacement of much of the Na with Al and Mg (Barshad, 1960; Bar-On and Shainberg, 1970). The Na-montmorillonite prepared by Xu (1988), on the other hand, was always washed in pH ≃ 6 solutions and stored in 0.01 *M* NaCl before washing with solutions of the desired electrolyte concentration. Miller (1984) claims that, because he observed a constant μ over the pH range 5.7 to 7.8, his preparations are free of hydrolyzed Al; however, Harsh et al. (1988a) have shown that μ for an OH–Al–smectite is relatively insensitive to pH in this region (Fig. 18, Section III.A.2).

Recently, yet another study has shown that the ζ-potential for Na–, Li–, and Cs–(homoionic)montmorillonites and illites was constant over a range of electrolyte concentration from 0.03 to 0.3 *M* of the accompanying Cl salt (Horikawa

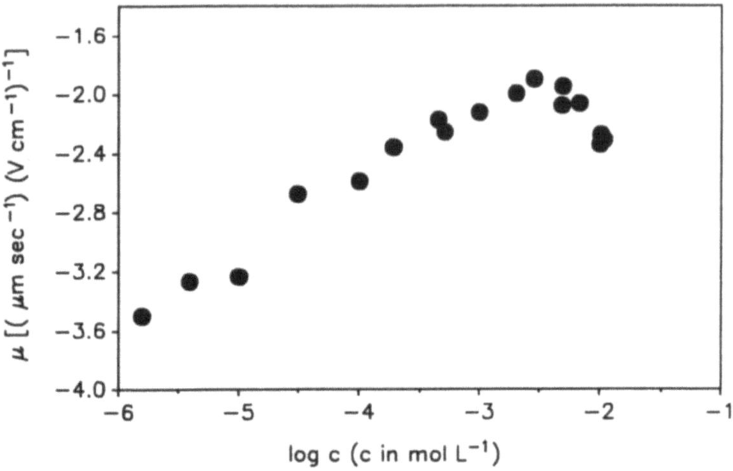

Figure 4. The electrophoretic mobility (μ) of Na–(Arizona) montmorillonite as a function of log concentration of NaCl. (Adapted from Xu, 1988.)

et al., 1988). The authors washed the clays with acidified molar NaCl, followed with molar neutral NaCl to obtain a pH of 6, and washed five times with molar Na or Li or 0.1 *M* Cs to saturate with the desired cation. They then removed excess electrolyte with methanol to avoid hydrolysis, and dried it at 25°. The dried clay was resuspended in deionized water by sonication, suspended in the desired electrolyte solution at a clay concentration of 0.01%, and shaken at 20° for 4 hours. This is essentially the same method of preparation used by Swartzen-Allen and Matijević (1975) except that the latter authors resuspended dried Na- or Cs-montmorillonite in distilled water at a clay concentration of 0.1% and did not state the time of storage before they determined μ. Swartzen-Allen and Matijevic (1975), even in the more concentrated suspension, found that nearly 100 and 50% of the Na and Cs, respectively, were desorbed from the clay, presumably replaced by Mg and Al.

The data of Xu (1988) for Na–(Arizona)montmorillonite show that μ was linearly related to the concentration of NaCl from 10^{-6} to 5×10^{-3} *M*. Between 10^{-3} and 10^{-2} *M* NaCl, μ reached a maximum and declined as the concentration was further increased (Fig. 4). This trend is consistent with that of Callaghan and Ottewill's (1974) data and contrasts with Miller's results, which showed no dependence of μ on concentration. Xu (1988) found that the same basic trend shown for Na–(Arizona)montmorillonite was followed for the Wyoming montmorillonite and Llano, Texas, vermiculite, regardless of the monovalent ion (Li, Na, K, Rb, Cs, Ag, Tl) used. Also, in contrast to other studies (Mattson, 1929ab; Lyons et al. 1981; Miller, 1984), Xu (1988) showed that μ was not directly related to the lyotropic series for alkali metals that generally describes their selectivity:

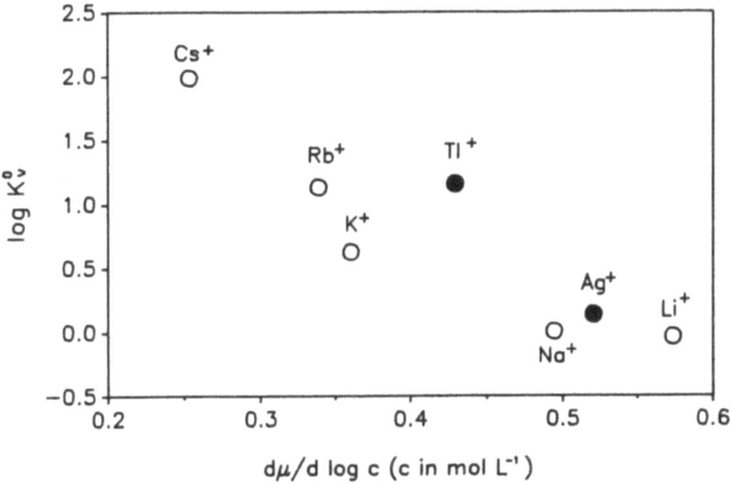

Figure 5. The relation between the Vanselow selectivity coefficient at one-half mole fraction for Na → M exchange and the rate of change of mobility with electrolyte concentration. (Adapted from Xu, 1988.)

$$\text{Li} < \text{Na} < \text{K} < \text{Rb} < \text{Cs}$$

but, in fact, the order of declining μ was highly concentration dependent. He did find that the rate of change of μ with the log of concentration [$d\mu/d$(log c)] was negatively related to alkali cation selectivity for the surface as defined by the Vanselow selectivity coefficient at 0.5 mole fraction (Fig. 5). Apparently, $d\mu/d$(log c) is related to the affinity of alkali cations for the surface, possibly as a measure of their ability to transfer across the shear plane in response to a concentration gradient.

2. Determination of ζ from μ

Extensive calculations of ζ for 2:1 clay minerals from μ using the Smoluchowski equation have been carried out by Callaghan and Ottewill (1974), Delgado et al. (1985), Friend and Hunter (1970), Mattson (1929a), Miller (1984), Low (1958, 1981), Pashley (1985), Ravina and Zaslavsky (1968), Swartzen-Allen and Matijević (1975), van Olphen (1957), and Xu (1988). In addition, Lyons et al. (1981) calculated ζ from streaming potential measurements of freshly cleaved mica sheets. A wide range of ζ values and inconsistent dependence on concentration is again indicative of the strong influence of procedure on the values obtained. All methods that included repeated washing or dialysis in distilled or deionized water (Callaghan and Ottewill, 1974; Delgado et al., 1985; Low, 1958; Mattson, 1929a; Miller, 1984) resulted in constant or irregular ζ response to concentration changes; however, when contact between the saturated clays and dilute solutions

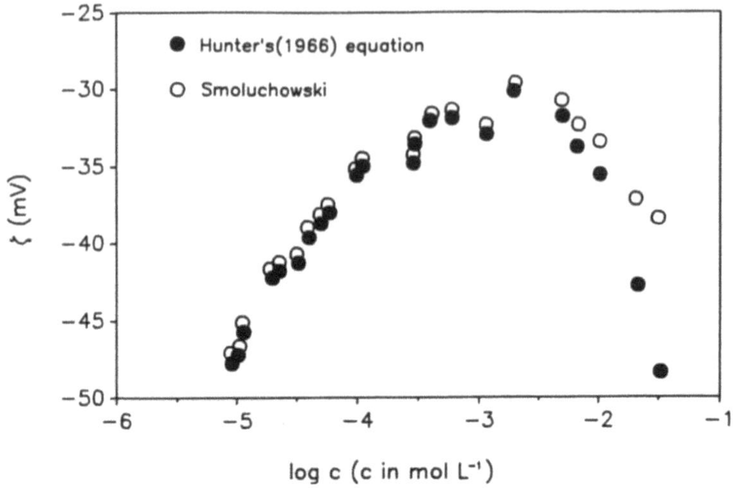

Figure 6. A comparison between Hunter's (1966) equation, correcting for nonconstant ε and η, and the Smoluchowski equation. (Adapted from Xu, 1988.)

was minimized, ζ became more positive with increasing concentration up to ~ 10^{-2} *M* (Lyons et al., 1981; Pashley, 1985; Xu, 1988). This suggests, in agreement with Lyons et al., that the presence of Mg^{2+} and hydrolyzed Al on the surface was responsible for the constant potential behavior observed in clays treated by extensive washing or dialysis with deionized water.

As discussed previously, application of the Smoluchowski equation requires that $\kappa a \gg 1$ and that *D* and η are everywhere constant and equal to their values in bulk solution. Regarding the latter assumption, Low (1976) provided evidence that η is not constant within the interfacial region of a charged clay particle and suggests that *D* also varies (Low, 1981). Hunter (1966) constructed graphs relating η and *D* to $d\psi/dx$ in the vicinity of a charged particle based on the equations of Booth (1951).

$$D = D_o [1 - B(d\psi/dx)^2] \quad (6)$$

and Lyklema and Overbeek (1961)

$$\eta = \eta_o [1 + f(d\psi/dx)^2] \quad (7)$$

where the subscript "o" denotes the values for bulk water and *B* and *f* are constants. Hunter (1966) estimated that $B = 1\text{-}12 \times 10^{-18}$ m² v⁻² (Low, 1981) and Hunter and Leyendekkers (1978) found that $f \simeq 10^{-15}$ $m^2\ v^{-2}$ in the vicinity of clay mineral surfaces. Low (1981) applied this correction to calculate ζ for Na-montmorillonite in 10^{-4} *M* NaCl and found that it caused less than a 1% deviation from the Smoluchowski equation. Xu (1988) compared the results of the Hunter (1966) correction and the uncorrected Smoluchowski equation and found significant deviation only at high concentration (Fig. 6).

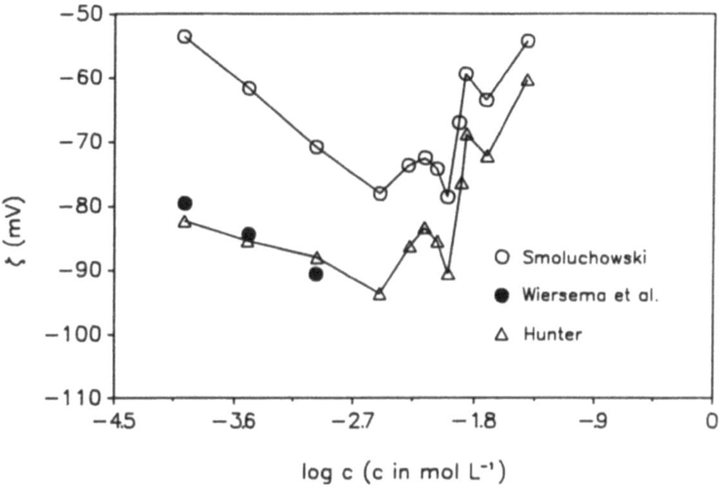

Figure 7. A comparison among Hunter's (1962) equation, correcting for relaxation and retardation effects, the complete analytical solution of Wiersema et al. (1966), and the Smoluchowski equation for calculation of ζ of Li-vermiculite. (Adapted from Friend and Hunter, 1970, with permission.)

The first assumption of the Smoluchowski equation ($\kappa a \gg 1$) also may not apply to smectite particles as a result of their finite size, their irregular surface topography in suspension (Pickles and Schlup, 1985; Shomer and Mingelgrin, 1978), and the possibility that they rotate with respect to the electric field. This has led many investigators to consider clay particles as spheres with radii estimated from the average particle dimension as deduced (1) from electron micrographs (Delgado et al., 1985), (2) from a desired fit to experimental data (Pashley, 1985), or (3) from particle settling velocities (Friend and Hunter, 1970; Xu, 1988). This allows the use of O'Brien and White's (1978) approach, which takes into account relaxation, retardation, surface conductance, and nonconstant η and D under a myriad of experimental conditions.

Using a computer solution (Wiersema et al., 1966) to correct for the preceding effects, Friend and Hunter (1970) were the first to report the effect of using lower values of κa to calculate ζ of Li-vermiculite (Fig. 7). Because relaxation and retardation slow the particle down, it is not surprising that the corrected ζ had a larger magnitude than when it is calculated by the Smoluchowski equation. The qualitative trend in the data, however, was not changed. Essentially the same correspondence between the Smoluchowski equation and the O'Brien and White (1978) approach were obtained by Delgado et al. (1985) for Na-montmorillonite and by Xu (1988) (using the analytical expression developed by Ohshima et al., 1983, which adapts the O'Brien and White approach for any counterion of known charge and mobility) for Wyoming montmorillonite, Arizona montmorillonite, and Llano, Texas, vermiculite saturated with the monovalent cations stated

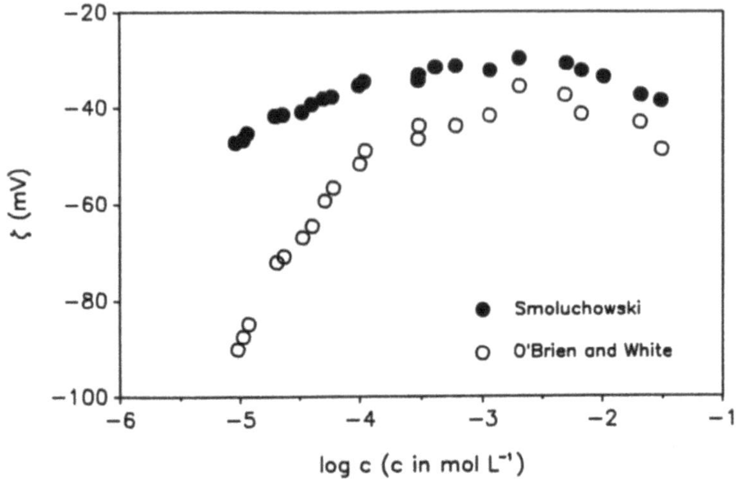

Figure 8. A comparison between the Smoluchowski equation and O'Brien and White's approach for calculating ζ for Li-(Wyoming)montmorillonite. (Adapted from Xu, 1988.)

earlier. Figure 8 compares the two approaches for Li-SWy-1 (Clay Minerals Society Source clay designation).

As stated previously, one result of the O'Brien and White approach is that $|\mu|$ is shown to have a maximum value when $10 < \kappa a < 100$, which is the range expected for ~0.3 μm spherical particles in electrolyte solutions between 10^{-4} and 10^{-2} (see Fig. 3). It follows that there is not a unique solution for ζ for a given value of μ. The choice between the high and low ζ for a given μ is, therefore, arbitrary. Delgado et al. (1985) and Pashley (1985) chose the lower of the two values, whereas, Xu (1988) calculated ζ from both. When he used the higher ζ, a minimum occurred around 10^{-3} *M* (Fig. 9) in place of the maximum seen in Figure 8, where the lower ζ was used. Hunter (1981) has suggested that the choice of the higher ζ may help to explain the anomalously low values obtained for particle of high intrinsic charge density, such as clay minerals. We will return to this discussion in the section on charge density of the shear plane.

3. Relation between ζ and ψ_δ

Given the uncertainty surrounding the calculation of ζ, one may wonder why all the concern is given to knowing this particular interfacial potential that is apparently significant only in electrokinetic studies. The interest stems from the idea that, at least in some systems, the shear plane is coincident with the outer Helmholtz plane (OHP) (Lyklema, 1977; Eversole and Boardman, 1941; Hunter and Alexander, 1963; Bagchi, 1979; Miller, 1984; Low, 1987). Some of the evidence comes from the application of the following equation, developed by Eversole and Boardman (1941):

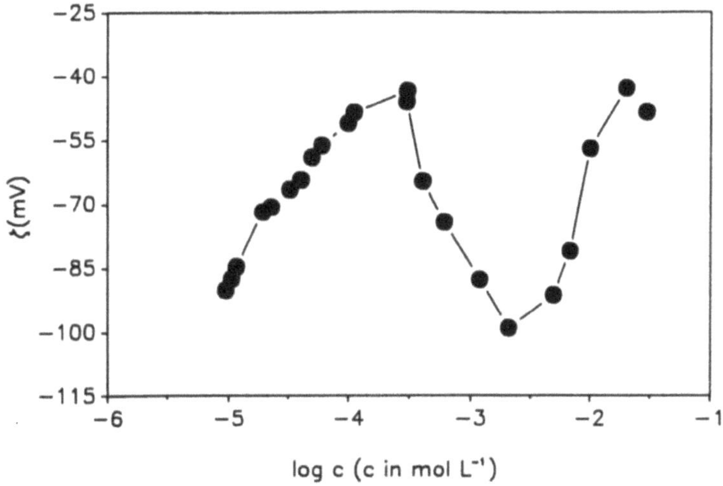

Figure 9. The ζ-potential calculated using O'Brien and White's approach and selecting ζ from the "high side" of the E versus y curve of Figure 3. (Adapted from Xu, 1988.)

$$\ln \tanh\left(\frac{ZF\zeta}{4RT}\right) = \tanh\left(\frac{ZF\psi_\delta}{4RT}\right) - \kappa\tau \tag{8}$$

This equation is derived by integrating the Poisson–Boltzmann equation (Eq. 1) and assuming that (1) the bathing electrolyte is symmetrical and (2) that $\psi(x) \ll 100/Z$ mv at all values of x within the boundary conditions of the integration (Hunter, 1981). If the Poisson–Boltzmann does apply to this region, a plot of ln tanh($ZF\zeta/4RT$) against κ should give a straight line from which ψ_δ (the potential at the OHP) and τ (the distance between the shear plane and OHP) can be obtained from the intercept and slope, respectively.

For oxide surfaces (glass and "ceramic") Eversole and Boardman (1941) (as reported by Hunter, 1981, p. 211) were able to obtain linear plots of Equation 8, indicating that the Poisson–Boltzmann equation was valid for the interparticle region. They obtained values for τ from 16 to 49 Å depending on the material and method used to determine ζ. Hunter and Alexander (1963), on the other hand, found τ = 10 Å and $\psi_\delta = -55$ mV for Na-kaolinite. Because kaolinite is generally regarded as a variable-charge mineral with little or no permanent charge, it is expected that its behavior should show constant potential behavior (with respect to electrolyte concentration) like the oxides.

Miller (1984) used his ζ values for Otay, Texas, and Upton montmorillonites to prepare plots from Equation 8 and obtained linear plots with $\tau = 0$ and ψ_δ between −49 and −71 mV depending on the mineral and saturating cation. Examples are shown for Li–, Na–, and K-Upton montmorillonite in Figure 10. The linear plots were expected since Miller had already found ζ independent of electrolyte concentration. Xu (1988), having shown marked concentration

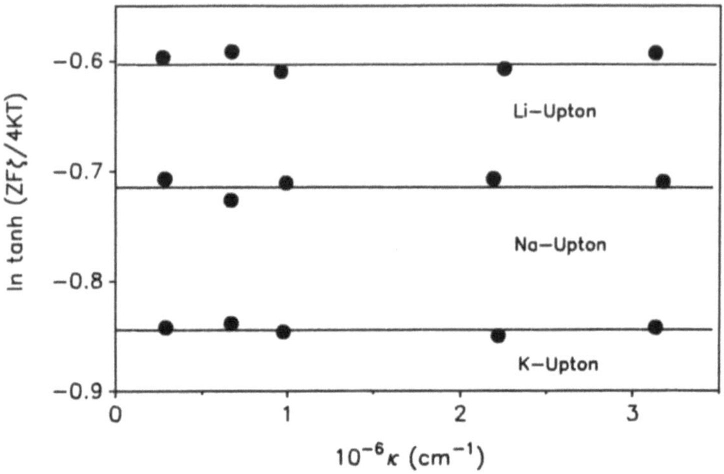

Figure 10. A plot of ln tanh($ZF\zeta/RT$) versus κ to determine τ and ψ_δ for three Upton, Wyoming, montmorillonites. (Adapted from Miller, 1984.)

dependence, could not obtain linear plots of Equation 8. Again, the difference may well be due to differences in the way the clays were prepared.

A more serious criticism of the application of Equation 8, however, is that it is likely to break down in the double layer region near a clay mineral surface. As we have pointed out, it is appropriate to apply the Stern model to the interfacial region near the surface. Sposito (1984, p. 159) showed that the Poisson–Boltzmann equation fails to predict realistic values of the charge density at a smectite surface. Part of this problem arises from the assumption that the counterions exist as point charges. The problem is partially resolved, at least for 1:1 electrolytes at $c^0 < 0.1\ M$, by employing a modified Poisson–Boltzmann model in which finite ion radii are invoked to restrict their distance of closest approach. This modified approach has been reviewed recently by Carnie and Torrie (1984), but no related modification of Equation 8 to take account of the so-called compact layer has been made (Hunter, 1981, p. 211).

4. Determination of σ_ζ

As discussed in Section II.B.3, the accurate calculation of σ_ζ requires (1) that the model for calculating ζ from μ is valid, (2) that the rate of change of the electrostatic potential in the vicinity of a clay particle is described by the Poisson equation, and (3) that the distribution of ions outside the shear plane is described by the Boltzmann equation. These asumptions are implicit in Equation 5. To our knowledge, no alternative to Equation 5 exists for calculating σ_ζ from ζ. In this section we give examples from the literature of σ_ζ calculated from ζ, show the effect of the model used to determine ζ, and present the various conceptual models used to explain the observed behavior.

It was mentioned in Section II.B.2 that many consider ζ to be anomalously low compared to the very high charge density at the surfaces of 2:1 clay minerals (deduced from cation exchange capacities). This "anomaly" is particularly apparent when the charge density at the shear plane surface (σ_ζ) and that at the mineral surface (σ_0) are compared. Low (1981) used Equation 5, after calculating ζ from the Smoluchowski equation (Eq. 4), to determine σ_ζ for 34 Na–montmorillonites in 10^{-4} *M* NaCl. The ratio of σ_ζ to σ_0 ranged from 0.008 to 0.020. In other words, 98% or more of the surface charge was balanced by the net positive charge within the shear plane. Low (1981) further argued that τ must be $\simeq 0$ because (1) the viscous drag on the water from surface oxygens and immobilized cations would preclude shearing water inside the OHP plane and (2) viscous flow of water is known to occur between Na–montmorillonite sheets separated by as little as 1.2 to 1.3 nm. Therefore, Low concluded that >98% of the exchangeable Na on montmormillonite is within the OHP.

Similar results indicating that a very small fraction of cations resides in the diffuse layer have been obtained by Miller (1984) for montmorillonites, Hunter and Alexander (1963) for kaolinite, and Chan et al. (1984) for illite. All were initially Na-saturated and suspended in 10^{-4} *M* NaCl. Chan et al. (1984) used negative adsorption data for their determination and Miller used microelectrophoresis, the moving boundary method, negative adsorption, and membrane potential measurements to obtain similar values of σ_ζ for three Na–montmorillonites. Sposito (1987) recently pointed out that the negative adsorption estimates of ψ_δ for Na–smectites are flawed by the failure to consider that about one-half of the particles in dilute salt solution exist in bilayers. This results in a significant underestimation of $|\zeta|$.

Mattson (1929a) used solution analyses in a dialysis cell containing Na–smectite in one chamber and NaCl solution in the other to find from 5.35 to 20.9% of the cations in the diffuse layer as background electrolyte was raised from 5×10^{-3} to 10^{-1} *M*. Miller (1984) has pointed out, however, that these determinations can be regarded as qualitative only because of the need to use average concentrations of Na and Cl in each cell and to assume ideal solution behavior. Membrane potential measurements are also suspect because of the suspension effect.

Other workers besides Mattson (1929a) found that σ_ζ tends to become more negative as electrolyte concentration increases (Delgado et al., 1985; Friend and Hunter, 1970; Miller, 1984; Xu, 1988). This decrease occurred whether ζ was independent of electrolyte concentration or not. In fact, it can be seen by comparing Figure 11 with Figure 8 that σ_ζ is very insensitive to ζ—it is much more dependent on electrolyte concentration. Three hypotheses have been advanced to explain the observed σ_ζ versus *c* behavior, which we will refer to as (1) the constant potential hypothesis, (2) the ion exchange hypothesis, and (3) the co-ion adsorption hypothesis.

The constant potential hypothesis was expressed first by Mattson (1929b) and later advanced by Low (1987) and Miller (1984). As Low (1981) shows in Figure 12, if ψ_δ ($\psi_\delta = \zeta$ is assumed) is constant, an increase in κ must result in an increase in σ_ζ (see also Eq. 5). Mattson (1929b) suggested that there existed a

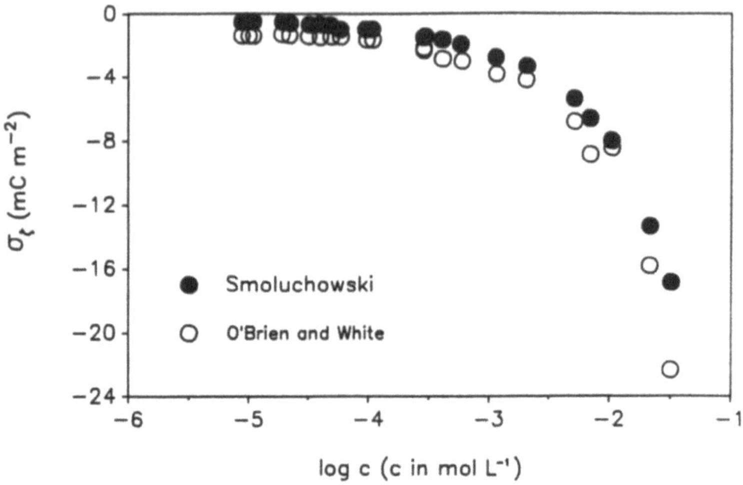

Figure 11. The surface charge density of Li–(Wyoming)montmorillonite at the shear plane as a function of log c when ζ is calculated from the Smoluchowski equation and from O'Brien and White's approach (the latter from the low side ζ of Fig. 3). (Adapted from Xu, 1988.)

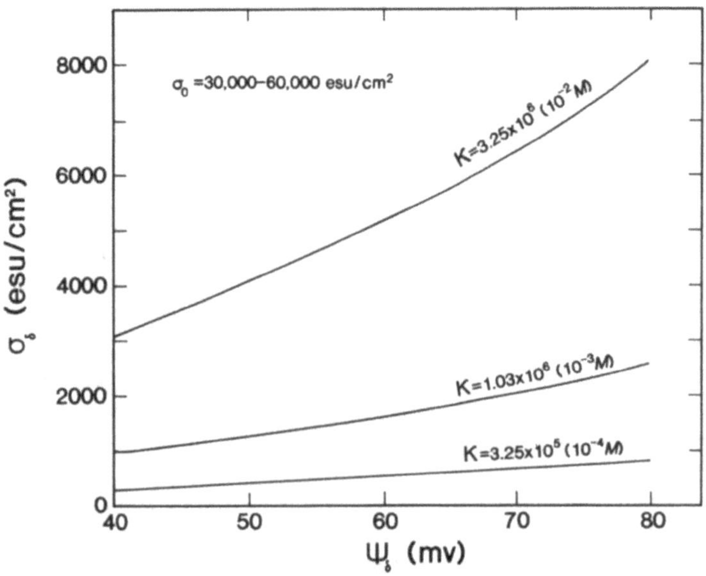

Figure 12. The effect of κ on the relation between σ_δ and ψ_δ. (Adapted from Low, 1987, with permission).

critical zeta potential above which cations were drawn into the Stern layer and below which dissociation occurred. This hypothesis depends on the several data sets that indicate that a constant potential exists at the shear (or Stern) plane, but, as pointed out in Section II.B.2, the observation of a constant ζ may be an artifact of sample preparation. It is also inconsistent with the mass action principle in that it requires that counterions move from the Stern layer to the diffuse layer as their bulk concentration increases.

The ion exchange hypothesis was suggested by Pashley (1981) to explain an increase in the magnitude of ψ_δ with concentration determined from repulsive force measurements on micas. It has also been applied to ζ determinations on 2:1 clays by Delgado et al. (1985) and on latex particles by Goff and Luner (1984). The hypothesis considers that competition for exchange sites by H ions at low Na concentration decreases the magnitude of σ_ζ because the H ions have a higher packing density at the surface. As Na concentration increases, H^+ is replaced and $|\sigma_\zeta|$ increases. Xu (1988) pointed out that exchangeable Al and Mg, are released from the clay lattice at low electrolyte concentration and will be replaced by Na^+ as [NaCl] increases. This latter exchange reaction has been shown to increase μ for OH–Al–hectorite (Harsh et al., 1988a).

Unlike the two previous hypotheses, which attribute the increase in negative-charge density to the behavior of the counterions, the co-ion hypothesis (Ma et al., 1981; Goff and Luner, 1984) attributes the increasing negative charge to an increase in co-ion adsorption relative to counterion adsorption. This hypothesis is not inconsistent with the mass action principle but may at first seem to be at odds with the concept of negative adsorption. This is not the case because the only requirement is that a *net* increase in negative charge occur. This can be brought about if anions are adsorbed to a greater extent than cations above some threshold bulk solution concentration. Monte Carlo simulations (Carnie and Torrie, 1984), which assume the ions have finite size limiting their approach to the surface, show that the classical Poisson–Boltzmann equation underestimates anion concentration near a negatively charged surface as concentration increases above $\sim 0.1\ M$. In the presence of divalent cations, the underestimation of anion adsorption is highly significant (reported in Sposito, 1984, p. 160).

Recently, Midmore and Hunter (1988) have presented a case against co-ion adsorption as the cause of a maximum in plots of ζ versus c for monodisperse polystyrene latices. When ζ is determined in aqueous solutions of 10^{-4} to 5×10^{-2} KF, KCl, KBr, and KI, there is no significant difference between the curves for the different anions. This implies that all four anions are adsorbed to the same extent, which Midmore and Hunter think unlikely, or the co-ion adsorption hypothesis is invalid for explaining the ζ maximum for polystyrene latices.

A fourth hypothesis could be based on a variable position of the shear plane relative to the OHP. If the shear plane were to move away from the OHP in the direction of the bulk solution, this would result in a reduction in the magnitude of ζ and, presumably, σ_ζ. Unfortunately, the position of the shear plane cannot be independently verified, that is, its position is always model dependent. As a result, no evidence either supports or refutes this hypothesis aside from the application of Equation 8, which requires a constant ζ.

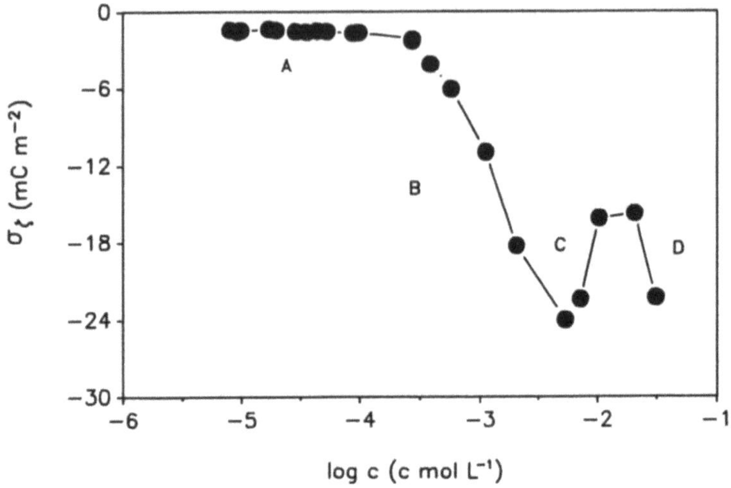

Figure 13. The surface charge density of Li–(Wyoming)montmorillonite at the shear plane as a function of log c when ζ is selected from the high side of Figure 3. (Adapted from Xu, 1988.)

The previously mentioned study of Midmore and Hunter (1988) is again germaine. They used high-frequency dielectric response measurements to determine the diffuse layer charge about a latex particle. This required an assumption that the technique measured only the conductivity of the ions in the diffuse layer and not the Stern layer. This assumption appeared to be supported by the fact that the calculated charge was only 3% of the titratable charge of the latex. They also used the high-frequency dielectric response measurements to calculate ζ using a model that required a measurement of surface conductance. Because the effect of ionic conductance inside the plane of shear on ζ was found not to be negligible, the model would appear to improve upon the O'Brien and White (1978) approach, which assumes that ionic conductivity is absent within the shear plane. A comparison of ζ and ψ_δ, then, yielded a deviation between them as the bulk concentration of ions decreased. Midmore and Hunter attributed this deviation to a movement of the shear plane away from the OHP, presumably brought about by a distortion of surface functional groups at the latex surface. The same argument would be difficult to apply to clay minerals because their surfaces generally appear to be too rigid to accommodate such distortion.

Midmore and Hunter's (1988) data are interesting, however, in that they also imply that the shear plane is not located inside the OHP. This would appear to validate the application of the Poisson–Boltzmann equation to the region located outside the shear plane. This, in turn, implies that Equation 5 may be used to estimate σ_ζ without invoking the Stern model or the modification of the Poisson–Boltzmann equation which restricts the closeness of counterion approach.

Xu (1988) has calculated σ_ζ from Equation 5 using ζ from Figure 9 and obtained a curve of σ_ζ versus log c_0, which contains four distinct regions (Fig. 13). Xu (1988) could explain this behavior by a model which essentially synthesizes the ion exchange and co-ion adsorption hypotheses. In region A, there is little change in σ_ζ in spite of the steep increase in ζ in the same region. In addition, the σ_ζ/σ_0 ratio here is very low in concert with other studies. This implies that there is very little dissociation of cations from the Stern layer at low NaCl concentration, which would be expected if the Stern layer were dominated by polyvalent cations like Al and Mg species. As NaCl concentration is increased (region B), Na hydrolysis decreases and exchangeable Mg and Al decrease. The clay becomes increasingly Na-saturated and, therefore, $|\sigma_\zeta|$ increases. It reaches a maximum between 10^{-3} and 10^{-2} *M* NaCl because the hydrolysis of Na-saturated clays is insignificant above $\sim 10^{-2}$ *M* NaCl (Shainberg, 1973).

The σ_ζ becomes more positive in region C as increasing [NaCl] in the presence of a nearly Na-saturated clay, drives diffuse layer Na^+ across the shear plane toward the surface in accordance with the mass action principle. The reversal in slope from region C to D, could occur if the "packing density" of Na^+ limited its transfer across the shear plane, while Cl^- adsorption continued. Thus, the complicated behavior shown in Figure 13 can be explained by a simple ion exchange and adsorption model that follows the dictates of the principle of mass action. Although its consistency with established chemical principles is, of course, an asset, no independent verification of the model exists. It is possible that Monte Carlo simulations could provide support, but at present no simulation of such a competitive system is available and would require careful analyses of exchanger composition at low electrolyte concentration. It must also be noted that the calculation of ζ and σ_ζ rest on unverifiable assumptions regarding particle geometry, the validity of the Poisson–Boltzmann equation, and the arbitrary choice of ζ from the "high" side of the μ versus ζ curve.

If all the positive charge inside the shear plane is assigned to cations that have formed surface complexes with the clay, the result can be compared to other estimates of Stern layer adsorption. Using a curve similar to Figure 13, but for Na-SWy-1 Xu (1988) calculated the maximum σ_ζ, which occurred at $\sim 5 \times 10^{-3}$ *M* NaCl. In this estimation, about 79% of the Na formed surface complexes. This value is significantly lower than Low's (1981) or Miller's (1984) estimates for similar systems (>98% of Na in surface complexes) mostly because of Xu's arbitrary choice of the "high side" ζ. The percentage of complexed Na was much larger than both the estimate of Shainberg and Kemper (1966) (36%) based on a theoretical Stern model approach, and an estimate by Sposito (1984, p. 34) of 23% based on negative adsorption data. Furthermore, Xu (1988) found no significant difference between the fractions of Li and Na in the Stern layer—a result consistent with his observation that there was no difference in selectivity between the two ions on montmorillonite, but inconsistent with the approximations of Shainberg and Kemper (1966) and Sposito (1984). The large discrepancy between these estimates underscores the need to develop a more comprehensive picture of the interfacial region.

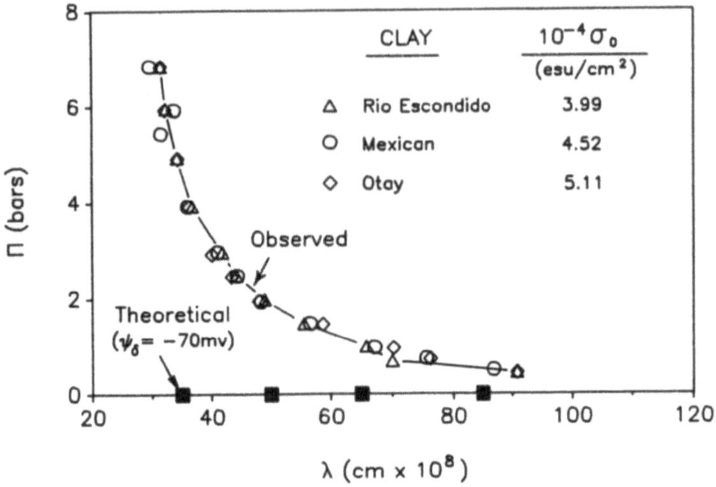

Figure 14. A comparison of theoretical and experimental curves of swelling pressure versus the distance between clay plates. (Adapted from Low, 1987, with permission.)

5. Correspondence between ζ and Repulsive Force Measurements

There are two views of the repulsive force between 2:1 clay minerals. The first ascribes the source of repulsion (or swelling pressure) to the osmotic pressure that develops from the diffuse layer cations. The other maintains that the osmotic contribution is relatively minor and that a solvation of hydration force, arising from the interaction between water molecules and the clay surface is primarily responsible for swelling. A lucid defense of the latter hypothesis, including a derivation of the following equation relating the swelling pressure to the distance between clay platelets, is given by Low (1981):

$$\Pi = p - f \tag{9}$$

where Π is the swelling pressure of the clay, p is the repulsive force between the clay plates, and f is the force due to van der Waal's attraction. The repulsive force, if due entirely to an osmotic gradient, depends on the potential at the midplane between the plates (ψ_h), the concentration of ions in the bulk solution, and the absolute temperature. The midplane potential must be determined as a function of the distance between the plates (λ) through the application of an appropriate model such as the diffuse layer model employing the Poisson–Boltzmann equation. The attractive force is a function only of λ.

Several studies have applied the diffuse layer model to clay swelling and found good agreement between the theoretical and experimental curves (Schofield, 1946; Norrish, 1954; Bolt and Miller, 1955; van Olphen, 1977; Quirk, 1968; Barclay and Ottewill, 1970). As Low (1981) has pointed out, all of these workers assumed that λ could be determined directly from the water content of the clay (i.e., all the water was located between the clay plates) and that the diffuse layer began at the clay surface. Viani et al. (1983) directly determined Π and λ for Na–montmorillonite using applied pressure simultaneously with low-angle x-ray diffraction to obtain the experimental curve in Figure 14. Low (1981) also applied the Stern model to determine ψ_h assuming (1) $\psi_\delta = \zeta \simeq -70$ mV, a value slightly more negative than determined by Miller (1984), and (2) a distance of 0.55 nm between the clay surface and the diffuse layer. The swelling pressures deduced by this analysis are insignificant relative the experimental values (Fig. 14). Low (1981) has concluded, therefore, that the swelling phenomenon derives primarily from hydration of the clay surface and that this hydration is independent of specific clay properties.

Pashley (1985) found a similar incongruity using direct force measurements between cleaved mica sheets and the calculated swelling pressure from ζ. He considered the possibility that the discrepancy arose from the assumptions implicit in the Smoluchowski equation and applied instead the O'Brien and White (1978) approach using κa as a fitting parameter. When $a = 0.25$ µm was chosen as the effective spherical radius, ζ ranged from -40 mV in 10^{-2} *M* KCl to -85 mV in 10^{-1} *M* KCl. These ζ potentials, in turn, gave good agreement between the force determinations and Stern model calculation of the repulsive force. Pashley expressed discomfort, however, in the use of a as a fitting parameter in the absence of a true description of the effect of geometry on the μ verus ζ relation. Pashley (1981) concluded in earlier studies that the repulsive force is, in part, a consequence of "hydration forces," but claims that these forces arise from the desolvation of adsorbed cations, because the magnitude and onset of the hydration force depend on the nature of the counterion. Sposito (1987) also recalculated ζ from literature values of μ for Na–smectite. He used the O'Brien and White (1978) approach and assumed that $\kappa a \simeq 10$ for smectite particles with dimensions of 0.3 µm $\times$ 0.3 µm $\times$ 1 nm. He obtained ζ of the order of -130 to -150 mV, which are significantly higher than other reported values. He used this exercise only to demonstrate the significant effect relaxation may have on μ, but stressed that it was just an approximation.

This last example serves as a fitting statement of the problem confronting those who would use microelectrophoresis to develop quantitative descriptions of interfacial phenomena such as ion adsorption and clay swelling. Such a description is not possible until two fundamental issues are resolved. First, the location of the diffuse double layer relative to the shear plane must be established, especially if it is found that ζ varies with the surface potential. Evidence for the coincidence of the OHP and shear plane is lacking and may never be unequivocally established. Second, equations based upon the geometry of a disc, such as the 2:1 clay minerals, must be developed to eliminate inadequate approximations using infinite, flat plates or "effective" spheres as models. If only the second task were established, it would take us a long way toward understanding the clay mineral–aqueous interface. The final distance will be covered only through a better understanding of solvent effects on surface–ion interactions.

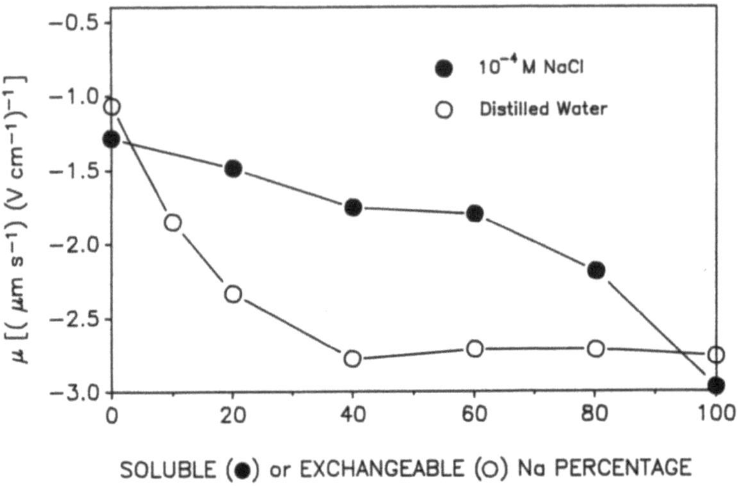

Figure 15. The electrophoretic mobility of Na,Ca–montmorillonite as a function of exchangeable sodium percentage in deionized water (Bar-On et al., 1970) or soluble Na charge percentage in deionized water (Bar-On et al., 1970) or soluble Na charge percentage in 10^{-4} *M* NaCl (Harsh and Hocking, unpublished data).

III. Qualitative Applications

We can learn much about the interfacial region of 2:1 clays in a qualitative sense by direct application of μ to surface phenomena. Because these applications do not require that μ be transformed to ζ and assume no theoretical model of the potential or ion distribution in the interfacial region, they are not confined by the assumptions implicit in the application of the Smoluchowski or Poisson–Boltzmann equations. As examples of these applications, in this section we present data related to demixing of Na and Ca on montmorillonite, the relationship between ion softness and selectivity on clays, and the structure of allophane and imogolite.

A. Application to the Nature of Adsorbed Cations

1. Demixing of Na and Ca on Montmorillonite

The electrophoretic mobility in deionized water of a mixed solid solution of Na- and Ca-montmorillonite as a function of the exchanger composition is shown in Figure 15 (Bar-On et al., 1970). As the Ca-montmorillonite is mixed with even small amounts of Na-montmorillonite the electrophoretic mobility increases dramatically. These results seem surprising particularly in light of the tendency for Ca-montmorillonites to form quasicrystals even in solutions with no added bulk electrolyte (Norrish and Quirk, 1954, reported in Sposito, 1984, p. 198). Quasicrystals of Ca-montmorillonite contain between four and eight particles at equilibrium (Sposito, 1984, p. 199); therefore, if the Na-montmorillonite were randomly mixed with the Ca-montmorillonite, approximately 20% of the Na would be located on the external surfaces of the quasicrystal and only a portion of this Na would reside outside the shear plane—that is, in the diffuse layer of the external surfaces.

Bar-On et al. (1970) concluded from the preceding observations that most of the Na must occupy external sites, suggesting a preference of Ca for interlayer sites. This is consistent with the tendency for Ca–montmorillonite to form quasicrystals in the first place, whereas Na–montmorillonite tends to exist in single sheets or bilayers in dilute solutions. The "demixing" continues to occur even after the quasicrystals begin to break up into smaller particles at ~30% exchangeable Na. At this point there is apparently sufficient Na to saturate external surfaces and no further increase in the magnitude of μ occurs.

The second curve on Figure 15 shows μ when the same experiment is performed at a total normality of 1.2×10^{-4} mole L^{-1} (Harsh and Hocking, unpublished data) and the x-axis gives the equivalent fraction of Na in solution ($\tilde{E}_{Na}$). The curve shows that at this electrolyte concentration, Ca dominates the montmorillonite's exchange sites even when $\tilde{E}_{Na} > 0.6$. This is confirmed by an estimate of 0.01 equivalent fraction Na on the montmorillonite when $\tilde{E}_{Na} = 0.8$, assuming a nonpreference isotherm (Sposito, 1981, p. 138). The calculated equivalent fraction overestimates the actual amount of exchangeable Na because montmorillonites show a significant preference for Ca relative to Na (Bruggenwert and Kamphorst, 1982, pp. 155–156). The curve supports the demixing phenomenon in that an increase in the magnitude of μ is apparent even at <1% Na on the exchanger. This would not appear likely if Na was not excluded from interlayer sites.

2. Smectites Containing Mixtures of Monovalent and Polyvalent Cations

In Section II of this chapter, considerable discussion was devoted to the effect of hydrolysis of Na-saturated clays on the ultimate composition of exchange sites. We pointed out that Na–clays were rapidly and extensively converted to M-clays, where M refers to Al, Mg, and other metals released from the octahedral layer of the solid (Shainberg, 1973). The effect of this transformation on μ is shown by some experiments performed on Chambers (Arizona) montmorillonite saturated with either Na or Cs (Swartzen-Allen and Matijević, 1975). Using radiotracers, they first showed that the Na- and Cs-saturated clays underwent ~100 and 50% exchange, respectively, with, presumably, Mg and Al upon dilution in doubly distilled water to 0.1 concentration. This further confirms the impossibility of maintaining a stable Na-saturated montmorillonite in dilute electrolyte.

The electrophoretic mobility of the "Na"– and "Cs"–montmorillonite as a function of pH is shown in Figure 16. It is interesting that there is very little pH-dependence below pH 8, indicating that Al hydrolysis and edge site charging have little effect on the shear plane potential in this range; whereas, when Al is added from solutions of 10^{-4} and 10^{-3} M $Al(NO_3)^3$, significant pH-dependence is observed (Fig. 17). There are several reasons that a completely hydrolyzed Na–clay might differ from the clay in $Al(NO_3)_3$ solution including (1) dominance of the exchange sites of the initially Na-saturated clay with Mg, which is more

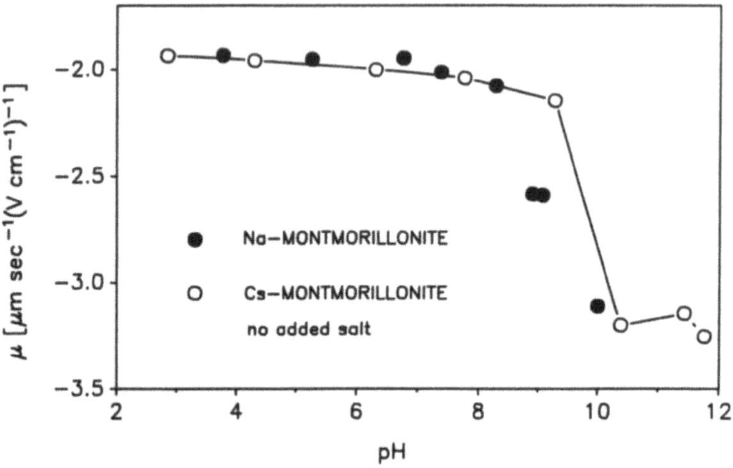

Figure 16. The change in electrophoretic mobility with pH for montmorillonites initially saturated with Na or Cs. (Adapted from Swartzen-Allen and Matijević, 1975, with permission.)

rapidly released than Al from octahedral sites of 2:1 clays and does not undergo hydrolysis in this region, (2) relegation of Al that is released and hydrolyzed to interlayer sites by a "demixing" phenomenon so that it does not affect μ as the pH is changed, and (3) a near-zero rate of change in edge site charge between pH 4 and 8.

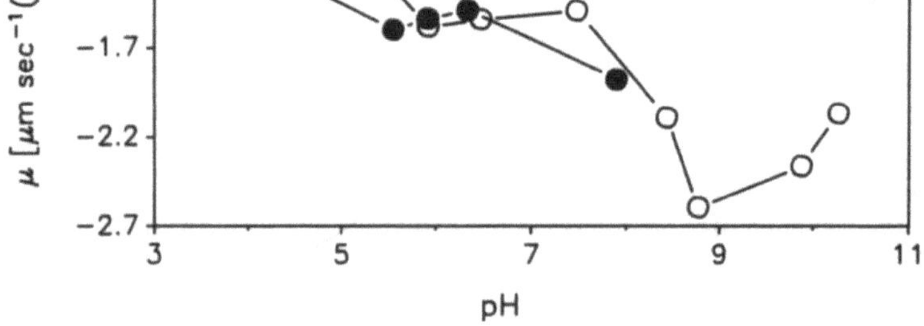

Figure 17. The electrophoretic mobility of initially Na-saturated montmorillonite as a function of pH in 10^{-3} and 10^{-4} M $Al(NO_3)_3$. (Adapted from Swartzen-Allen and Matijević, 1975, with permission.)

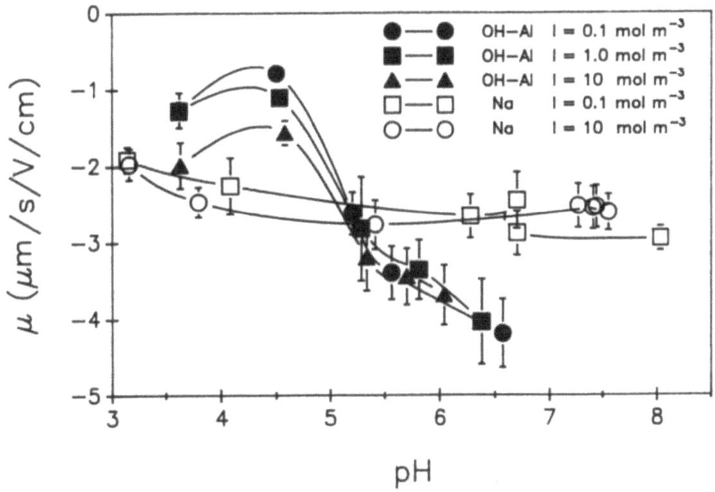

Figure 18. The electrophoretic mobility of OH-Al-hectorite and initially Na-saturated hectorite as a function of pH in 10^{-2}, 10^{-3}, and 10^{-4} *M* NaCl solutions (Harsh et al., 1988a).

Swartzen-Allen and Matijević (1975) attributed the rapid decrease in μ between pH 8 and 10 to charging of Al–OH groups at the edge sites of the octahedral layer. Although this was a likely contributor to the decrease, it is also plausible that Na–Mg exchange occurred, resulting from the high levels of NaOH that had to be added to achieve the high pH. Preferential Mg → Na exchange at external sites would allow small additions of Na to have a relatively large effect on μ.

The same authors also demonstrated that addition of $Al(NO_3)_3$ and $Th(NO_3)_4$ to Na–montmorillonite suspensions solutions at pH 3.5 (to preclude significant hydrolysis in solution) cause μ to increase to near zero for Al and to positive values for Th as the molarity of the polyvalent cation was increased to 10^{-3} and 2×10^{-3}. This contrasted with lyophobic colloids (AgI, various latexes, etc.), which show charge reversal only in the presence of hydrolyzed metal cations. They suggested that ion selectivity was the governing factor in a cation's effect on μ and that the increasing selectivity with cation charge accounted for differences between Al and Th. They alternatively proposed that hydrolysis was enhanced by the clay and that the hydrolysis products were preferentially sorbed but contributed less positive charge.

Harsh et al. (1988a) precipitated $Al_x(OH)_y^{3x-y}$ on Na–hectorite and determined μ as a function of pH at three different ionic strengths. Unlike the preparation of Na–clay in $Al(NO_3)_3$ solution (Swartzen-Allen and Matijević, 1975), the OH-Al-hectorite was washed with 1 *M* $NaClO_4$ to remove exchangeable Al and aged for several weeks. At $3.5 < pH < 5.5$, increasing $NaClO_4$ concentration from 10^{-4} to 10^{-2} *M* caused μ to become more negative (Fig. 18). We proposed that Na exchange for Mg and Al produced a more negative μ and confirmed that exchangeable Al and Mg were reduced as [NaCl] increased at the low pH values.

Above pH 5.5, μ was more negative than for an untreated Na–hectorite. The reason for this behavior is unclear but seems to indicate that the presence of $Al_x(OH)_y^{3x-y}$, which should impart positive charge to the mineral at low pH, did not directly affect μ. It may have indirectly affected cation distribution by causing quasicrystal formation which, in turn, led to an unequal distribution of Na and Al + Mg between external and interlayer sites, respectively. Harsh et al. (1988a) also found that μ became less negative with time at low pH on both Na- and OH–Al–hectorite, a result consistent with the hydrolysis and interchange reactions, which led to replacement of exchangeable Na with Mg from the octahedral layer or Al from the precipitated $Al_x(OH)_y^{3x-y}$.

3. Monovalent Cation Selectivity

As discussed in Section II.B.1 and 3, Xu (1988) did not find consistent correlations between the selectivity coefficient of a given monovalent cation (M) in exchange with Na^+ and its absolute μ at a given electrolyte concentration. This would appear to be a result of the fact that at low ionic strengths M-saturated clays are not stable and that at high ionic strengths complications such as double-layer compression, quasicrystal formation, and co-ion adsorption may interfere with unambiguous interpretations of μ. Xu (1988) did find a correlation between the rate of change of μ with concentration and the log of the Vanselow selectivity coefficient (K_v) determined at 0.5-mole fraction and a constant ionic strength of 0.01 *M*.

Given the apparent relationship between the ability of a cation to transfer across the shear plane in response to a concentration gradient and its affinity for the surface (see Fig. 5), Xu (1988) also attempted to correlate the surface charge calculated at 5×10^{-3} *M* MCl (using the approach described for the data in Fig. 13) with log K_v. The result was a negative linear correlation with $r^2 = 0.96$ for Wyoming montmorillonite (SWy-1). The implication of a relationship between the charge at the shear plane and a selectivity coefficient is that increasing affinity reflects an increasing tendency for the cation to form surface complexes with the mineral.

In accord with the idea that surface complexes influence cation selectivity on clay minerals, Sullivan (1977) and Sposito (1984, p. 129ff) applied the principle of hard and soft acids and bases (HSAB) to cation affinity. The HSAB principle states that Lewis acids classified as hard (i.e., tending to act as hard spheres and undergoing primarily electrostatic interactions) prefer to complex with hard Lewis bases, whereas soft (i.e., polarizable and likely to form covalent bonds) Lewis acids complex preferentially with soft Lewis bases. In an aqueous solution, softer cations are more likely to form inner-sphere complexes with any ligand softer than water, which is a very hard base.

Applied to clay minerals and monovalent cation selectivity the HSAB principle suggests that softer cations show a greater tendency to form inner-sphere complexes with surface oxygens of clay minerals if they are softer than water. If the ability to form such complexes determines the selectivity of a cation for a given

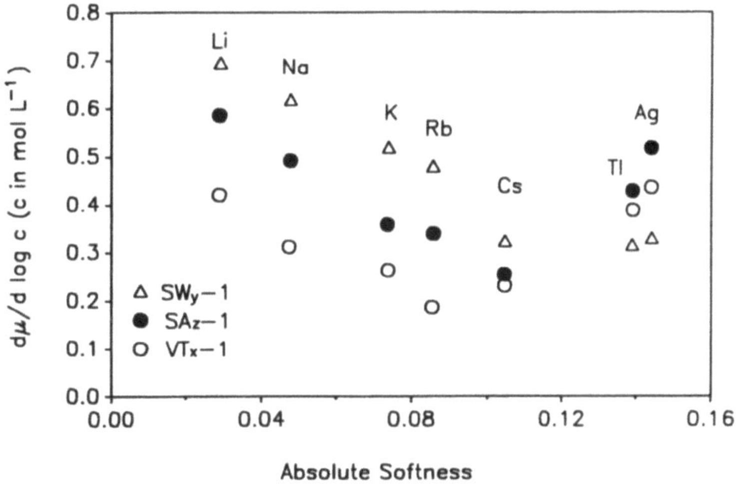

Figure 19. The relation between $d\mu/d\log c$ and the softness of monovalent cations on three clay minerals. (Adapted from Xu, 1988.)

mineral, a relationship should exist between $d\mu/d\log c$ and cation affinity. That is, if cations form inner-sphere complexes with surface oxygens, they should show less tendency to transfer across the shear plane in response to a concentration gradient. In agreement with this, a linear relationship is shown between cation softness and $d\mu/d\log c$ for Arizona montmorillonite (SAz-1), Wyoming montmorillonite (SWy-1) and Llano, Texas, vermiculite (VTx-1) (Fig. 19), but the linear relationship is interrupted by a minimum in $d\mu/d\log c$ that depends on the mineral. This may indicate that a degree of cation softness is reached beyond which complex formation with the generally hard surface oxygen bases is reduced. This point is reached at a lower softness on VTx-1 than on the montmorillonites; a finding consistent with the idea that the softness of the mineral surface decreases with increasing surface charge (Sposito, 1984, pp. 14ff).

B. Structural Models of Allophane and Imogolite

A microelectcrophoresis experiment provided support for proposed structural models of allophane (van Reeuwijk and de Villiers, 1968) and imogolite (Cradwick et al., 1972). In our laboratory, we have recently determined the points of zero net proton charge (PZNPC) and the points of zero charge (PZC) for three synthetic allophanes and imogolite by potentiometric titration and microelectrophoresis, respectively (Harsh et al., 1988b). The results of these determinations for imogolite and for the three allophanes (Al/Si = 1.12, 1.56, and 2.04) are presented in Figure 20. Points of zero net charge (PZNC) determined by Na^+ and Cl^- adsorption for similar materials from a study by Clark and McBride (1984) are also included.

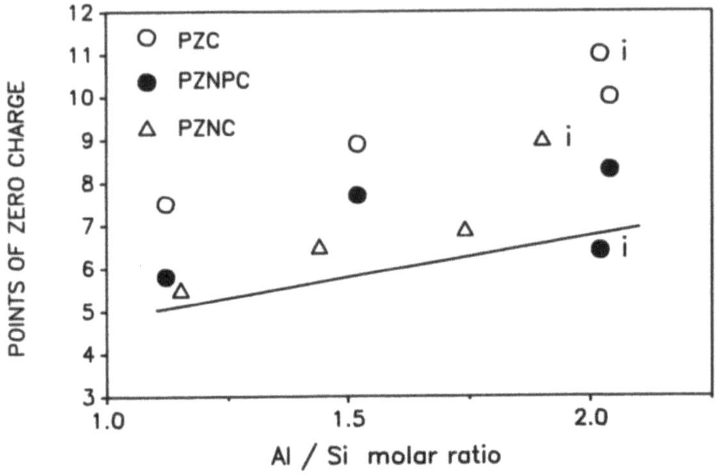

Figure 20. Points of zero charge for three allophanes and imogolite as a function of the Al/Si ratio of the parent solutions. (Adapted from Harsh, 1988b; and Clark and McBride, 1984.)

The PZNPC is a measure of the pH where the net surface charge due to titratable functional groups is zero. The PZNC, on the other hand, measures the pH at which total net charge due to both titratable groups and permanent structural charge (e.g., Al substitution in tetrahedral positions) vanishes. The PZC is the pH at which the net charge at the shear plane or, approximately, the OHP is zero. Thus, equivalence between the PZC and PZNPC implies that there is little or no permanent structural charge and equivalence between the PZC and PZNPC implies that the cation and anion form an equal number of inner- and outer-sphere complexes (Sposito, 1984, p. 81ff).

From these definitions we can interpret the results in Figure 20 to show first that there is very little structural charge in the 1.12 Al/Si allophane, because the PZC and PZNPC are nearly equivalent. This implies that the amount of tetrahedral Al occurring in the synthetic allophanes is small. This result is consistent with ^{27}Al NMR spectra of poorly ordered aluminosilicates, which show very low tetrahedral coordination of Al at Al/Si $>$ 1 in the solid phase (Wilson et al., 1986). The fact that the PZNPC and PZNC did not coincide for imogolite is surprising because NMR spectra show that there is not tetrahedrally coordinated Al in imogolite (Wilson et al., 1986). The difference, therefore, must have come from a failure of one of the methods to determine the respective point of zero charge. It is possible, for example, that equilibrium was not achieved in the ion adsorption measurements. If this were the case, the amount of Na^+ adsorption would be underestimated, increasing the PZNC. A similar failure for the 1.56 and 2.04 Al/Si allophanes is implied, because they also have Al/Si $>$ 1 for which tetrahedral substitution of Al for Si is unlikely (Wilson et al., 1986).

For all materials, the PZC was significantly higher than the PZNPC, implying that a net positive charge existed within the Stern layer when the net surface

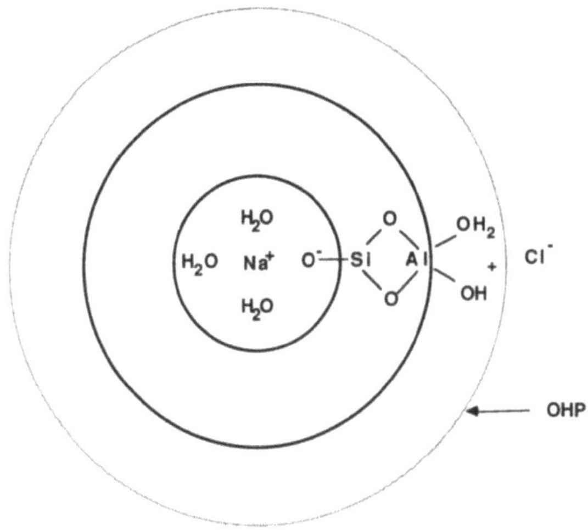

Figure 21. A diagrammatic representation of a cross-section of an imogolite tube or allophane spherule (Harsh et al., 1988b).

charge was zero. This can occur only if Na^+ forms a greater number of surface complexes than Cl^-. Both allophane and imogolite present a favorable environment for forming surface complexes with Na^+, if we accept the structural models proposed by Cradwick et al. (1972) for imogolite and by van Reeuwijk and de Villiers (1968) for allophane. A schematic diagram of a cross-section of imogolite tubules or allophane spherules based on these models is shown in Figure 21. According to the models, Si–OH groups, which are negatively charged even in acidic environments, protrude into the interior of the particle. Assuming that at least a portion of the adsorbed Na^+ enters the particle interior, it will show a higher tendency to form outer-sphere complexes than Cl, which is attracted to positively charged Al–OH sites on external surfaces. As expected, the largest difference between the PZNPC and the PZC was in the imogolite where the open tubular structure should allow easier and greater entrance of Na^+ into interior sites.

IV. Conclusions

Results from microelectrophoresis experiments give qualitative support for Stern or surface complexation models of the solid–aqueous interfacial region of aluminosilicate clay minerals, but a quantitative description of this interface remains elusive. We have shown that ζ and σ_ζ depend on the model used to calculate these parameters from μ. The geometry-independent Smoluchowski equation gives values of ζ that are different in magnitude than those that derive from the conventional use of the O'Brien and White (1978) approach, when clay minerals are approximated by spheres with $\kappa a < 100$. If ζ is taken from the "high side" of

μ versus ζ curves generated by the O'Brien and White (1978) approach, both the magnitude of ζ and the shape of the ζ versus electrolyte concentration curve differ from that calculated from the Smoluchowski equation. Further disagreement is seen when related physicochemical parameters such as the repulsive force between clay particles and amount of Stern layer adsorption are determined by other methods. The importance of these parameters to the description of ion behavior in soils and swelling; of soil clays underscores the need for valid theoretical models of the clay–aqueous interface so that techniques such as microelectrophoresis realize their potential as powerful tools in both theoretical and applied research.

In the interim, microelectrophoresis is still valuable as a tool for obtaining a qualitative understanding of the interfacial region. It has been used, for example, to verify demixing of Na and Ca between interlayer and external exchange sites on montmorillonite. We know that most polyvalent metal ions, even when partially hydrolyzed, do not easily reverse the charge at the shear plane of permanent-charge clay minerals, unlike the same metals on variable-charge oxides. Microelectrophorésis has shown the complexity of the behavior of smectites with a surface precipitate of $Al(OH)_3$ and implicated ion exchange and demixing phenomena as determinants of this behavior. The hard and soft acid/base model, applied to ion selectivity by assuming softer cations can form inner-sphere complexes with surface oxygens of 2:1 clays, is supported by microelectrophoresis results. Finally, allophane and imogolite particles both demonstrated an anomalous difference between the PZC and PZSE that could be explained by adsorption of Na^+ at interior sites while Cl^- is relegated to external sites. This hypothesis is consistent with proposed structural models of these poorly crystalline materials.

Acknowledgments. We would like to thank Drs. James Kittrick, David Mulla, and Garrison Sposito for their reviews and many helpful suggestions. This material is based in part on work supported by the National Science Foundation under Grant EAR8720813.

References

Bagchi, P., B.V. Gray, and S.M. Birnbaum. 1979. Preparation of model poly(vinyl toluene) latices and characterization of their surface charge by titration and electrophoresis. J. Colloid Interface Sci. 69:502–508.

Barclay, L.M., and R.H. Ottewill. 1970. The measurement of forces between colloidal particles. Spec. Disc. Faraday Soc. 1:138–147.

Bar-On, P., and I. Shainberg. 1970. Hydrolysis and decomposition of Na-montmorillonite leached with distilled water. Soil Sci. 109:241–246.

Bar-On, P., I. Shainberg, and I. Michaeli. 1970. Electrophoretic of montmorillonite particles saturated with Na/Ca ions. J. Colloid Interface Sci. 33:471–472.

Barshad, I. 1960. The effect of the total chemical composition and crystal structure of soil minerals on the nature of the exchange cation in acidified clays and in naturally occurring acid soils. Int. Congr. Soil Sci., Trans. 7th (Madison, WI) II:435–444.

Barshad, I., and A.E. Foscolos. 1970. Factors affecting the rate of the interchange reaction of adsorbed H^+ on the 2:1 clay minerals. Soil Sci. 110:52–60.

Bolt, G.H., and R.D. Miller. 1955. Compression studies of illite suspensions. Soil Sci. Soc. Amer. Proc. 19:285–288.

Booth, F. 1951. The cataphoresis of spherical fluid droplets in electrolytes. J. Chem. Phys. 19:1331–1336.

Bruggenwert, M.G.M., and A. Kamphorst. 1982. Survey of experimental information on cation exchange in soil systems. In Bolt, G.H. and M.G.M. Bruggenwert (eds.). Soil Chemistry: Part B. Physicochemical Models. Elsevier, Amsterdam.

Callaghan, I.C., and R.H. Ottewill. 1974. Interparticle forces in montmorillonite gels. Faraday Disc. Chem. Soc. 57:110–118.

Carnie, S.L., and G.M. Torrie. 1984. The statistical mechanics of the electrical double layer. Adv. Chem. Phys. 56:141–253.

Chan, D.Y.C., R.M. Pashley, and J.P. Quirk. 1984. Surface potentials derived from co-ion exclusion measurements on monoionic montmorillonite and illite. Clays Clay Miner 32:131–138.

Clark, C.J., and M.B. McBride. 1984. Cation and anion retention by natural and synthetic allophane and imogolite. Clays Clay Miner. 32:291–299.

Cradwick, P.D.G., V.C. Farmer, J.D. Russell, C.R. Masson, K. Wada, and N. Yoshinaga. 1972. Imogolite, a hydrated aluminum silicate of tubular structure. Nature 240:187–189.

Delgado, A., F. Gonzalez-Caballero, and J.M. Bruque. 1985. On the zeta potential and surface charge density of montmorillonite in aqueous electrolyte solution. J. Colloid Interface Sci. 113:203–211.

Eversole, W.G., and W.W. Boardman. 1941. The effect of electrostatic forces on electrokinetic potentials. J. Chem. Phys. 9:798–801.

Friend, J.P., and R.J. Hunter. 1970. Vermiculite as a model system in the testing of double layer theory. Clays Clay Miner. 18:275–283.

Goff, J.R., and P. Luner. 1984. Measurement of colloid mobility by laser Doppler electrophoresis: The effect of salt concentration on particle mobility. J. Colloid Interface Sci. 99:468–483.

Harsh, J.B., H.E. Doner, and D.W. Fuerstenau. 1988a. Electrophoretic mobility of hydroxy-aluminum- and sodium-hectorite in aqueous solutions. Soil Sci. Soc. Amer. J. 52:1589–1592.

Harsh, J.B., Y. Yang, J. Boyle, and T. Murarik. 1988b. Surface complex formation between sodium and noncrystalline aluminosilicates. Agron. Abst. p. 198.

Horikawa, I., R.S. Murray, and J.P. Quirk. 1988. The effect of electrolyte concentration on the zeta potentials of homoionic montmorillonite and illite. Colloids Surf. 32: 181–195.

Hunter, R.J. 1962. The calculation of zeta potential from mobility measurements. J. Phys. Chem. 66:1367–1368.

Hunter, R.J. 1966. The interpretation of electrokinetic potentials. J. Colloid Interface Sci. 22:213–239.

Hunter, R.J. 1981. Zeta Potential in Colloid Science. Academic Press, New York.

Hunter, R.J., and A.E. Alexander. 1963. Surface properties and flow behavior of kaolinite. Part I: Electrophoretic mobility and stability of kaolinite sols. J. Colloid Sci. 18:820–832.

Hunter, R.J., and J.V. Leyendekkers. 1978. Viscoelectric coefficient for water. J. Chem. Soc. Faraday 1 74:450–455.

Low, P.F. 1958. Movement and equilibrium of water and soil systems as affected by soil-water forces. In *Water and Its Conduction by Soils*. pp. 55–64. Nat. Acad. Sci.–Natl. Research Council, Special Report 40, Highway Research Board, Washington, D.C.

Low, P.F. 1976. Viscosity of interlayer water in montmorillonite. Soil Sci. Soc. Amer. J. 44:667–676.
Low, P.F. 1981 The swelling of clay III: Dissociation of exchangeable cations. Soil Sci. Soc. Am. J. 45:1074–1078.
Low, P.F. 1987. The clay-water interface. Proc. Internatl. Clay Conf., Denver, 1985. pp. 247–256.
Lyklema, J. 1977. Water at interfaces: A colloid-chemical approach. J. Colloid Interface Sci. 58:242–250.
Lyklema, J., and J. Th. G. Overbeek. 1961. On the interpretation of electrokinetic potentials. J. Colloid Sci. 16:501–512.
Lyons, J.S., D.N. Fourlong, and T.W. Healy. 1981. The electrical double-layer properties of the mica (muscovite)-aqueous electrolyte interface. Aust. J. Chem. 34:1177–1187.
Ma, C.M., F.J. Micale, M.S. El-Aasser, and J.W. Vanderhoff. 1981. In D.R. Bassett and A.E. Hamielec (eds). *Emulsion Polymers and Emulsion Polymerization*. pp. 251–262. ACS Symposium Series 165. American Chem. Soc., Washington, D.C.
Mattson, S. 1929a. The laws of soil colloidal behavior I. Soil Sci 28:179–220.
Mattson, S. 1929b. The laws of soil colloidal behavior II. Soil Sci. 28:373–409.
Midmore, B.R., and R.J. Hunter. 1988. The effect of electrolyte concentration and co-ion type on the ζ-potential of polystyrene latices. J. Colloid Interface Sci. 122:521–529.
Miller, S.E. 1984. The characterization of the electrical double-layer of montmorillonite. Ph.D. thesis. Purdue University. West Lafayette, Indiana.
Norrish, J., and J.P. Quirk. 1954. Crystalline swelling of montmorillonite. Use of electrolytes to control swelling. Nature 173:255–256.
Norrish, K. 1954. The swelling of montmorillonite. Faraday Soc. Dis. 18:120–134.
O'Brien, R.W., and L.R. White. 1978. Electrophoretic mobility of a spherical colloidal particle. J. Chem. Soc. Faraday Trans. II. 74:1607–1626.
Ohshima, H., T.W. Healy, and L.R. White. 1983. Approximate analytic expressions for the electrophoretic mobility of spherical colloidal particles and the conductivity of their dilute suspension. J. Chem. Soc. Faraday Trans. II. 79:1613–1628.
Ottewill, R.H., and J.N. Shaw. 1972. Electrophoretic studies on polystyrene latices. J. Electroanal. Interfacial Chem. 37:133–142.
Pashley, R.M. 1981. DLVO and hydration forces between mica surfaces in Li^+, Na^+, K^+ and Cs^+ electrolyte solutions. A: Correlation of double-layer and hydration forces with surface exchange properties. J. Colloid Interf. Sci. 83:531–546.
Pashley, R.M. 1985. Electromobility of mica particles dispersed in aqueous solutions. Clays Clay Miner. 33:193–199.
Pickles, D.G., and J.P. Schlup. 1985. Particle association in smectite soils by transmission electron microscopy. Clays Clay Miner. 33:362–366.
Quirk, J.P. 1968. Particle interaction and soil swelling. Israel J. Chem. 6:213–234.
Ravina, I., and D. Zaslavsky. 1968. Non-linear electrokinetic phenomena Part II. Experiments with electrophoresis of clay particles. Soil Sci. 106:94–100.
Schofield, R.K. 1946. Ionic forces in thick films of liquid between charged surfaces. Trans. Faraday Soc. 42B:219–225.
Shainberg, I. 1973. Rate and mechanism of Na-montmorillonite hydrolysis in suspensions. Soil Sci. Soc. Amer. Proc. 38:689–694.
Shainberg, I., and W.D. Kemper. 1966. Hydration status of adsorbed cations. Soil Sci. Soc. Am. J. 43:651.
Shomer, I.H., and U. Mingelgrin. 1978. A direct procedure for determining the number

of plates in tactoids of smectites: The Na/Ca-montmorillonite case. Clays Clay Miner. 26:135–137.

Sposito, G. 1981. *The Thermodynamics of Soil Solutions*. Oxford University Press, New York.

Sposito, G. 1984. *The Surface Chemistry of Soils*. Oxford University Press, New York.

Sposito, G. 1987. The ion distribution in a 1:1 electrolyte solution near a smectite surface. EOS Trans., Amer. Geophys. Union 68:1281–1282.

Stern, O. 1924. Zur Theorie der elektrolytischen Doppelschicht. Z. Elektrochem. 30:509–527.

Stigter, D. 1978. Electrophoresis of highly charged colloidal cylinders in univalent salt solution. J. Phys. Chem. 82:1417–1429.

Sullivan, P.J. 1977. The principle of hard and soft acids and bases as applied to exchangeable cation selectivity in soils. Soil Sci. 124:117–121.

Swartzen-Allen, S.L., and E. Matijevic. 1975. Colloid and surface properties of clay suspensions II: Electrophoresis and cation adsorption of montmorillonite. J. Colloid Interf. Sci. 50:143–153.

van Olphen, H. 1957. Surface conductance of various ion forms of bentonite in water and the electrical double layer. J. Phys. Chem. 61:1276–1286.

van Olphen, H. 1977. An Introduction to Clay Colloid Chemistry. Wiley–Interscience, London.

van Reeuwijk, L.P., and J.M. de Villiers. 1968. Potassium fixation by amorphous aluminosilicate gels. Soil Sci. Soc. Amer. Proc. 32:238–240.

Viani, B.E., P.F. Low, and C.B. Roth. 1983. Direct measurement of the relation between interlayer force and interlayer distance in the swelling of montmorillonite. J. Colloid Interface Sci. 96:229–244.

Wiersema, P.H., A.L. Loeb, and J.Th.G. Overbeek. 1966. Calculation of the electrophoretic mobility of a spherical colloid particle. J. Colloid Interface Sci. 22:78–99.

Wilson, M.A., S.A. McCarthy, and P.M. Fredericks. 1986. Structure of poorly-ordered aluminosilicates. Clay Miner 21:879–897.

Xu, Shihe. 1988. Electrophoretic mobility and monovalent cation selectivity of three reference clay minerals. M.S. Thesis. Dept. Agronomy and Soils. Washington State University. Pullman, Wash.

Applications of Luminescence Spectroscopy to Studies of Colloid–Solution Interfaces

S.J. Traina

I. Introduction

There has been a growing interest in the fields of colloid chemistry, surface chemistry, and soil chemistry, in the use of physical methods to study chemical reactions at colloid–solution interfaces. Ideally these methods should be noninvasive, sensitive, and require minimal sample pretreatment. Luminescence spectroscopy meets these criteria and has been used extensively in studies of both organic and inorganic colloidal systems. Typically, luminescent probe molecules are introduced at low concentrations into interfacial regions of interest. These molecules

Advances in Soil Science, Volume 14

can then report back to the spectroscopist on the nature of their surrounding chemical environment. Such experiments have provided detailed information on interfacial solvent polarities, ground-state and excited-state complex formation on colloid surfaces, molecular distributions of adsorbed species, and the configurations and relative mobilities of molecules in interfacial regions. One can obtain this wealth of information from many synthetic and natural systems of interest.

Extensive discussions of the applications of luminescence spectroscopy to colloidal and micellular interfaces can be found in earlier reviews (Thomas, 1984; 1987). These publications cover many aspects of fluorescence and spectroscopy in colloidal systems. The present discussion is more limited in scope, but it focuses on the use of organic fluorescent probe molecules to study model colloidal systems indicative of those commonly found in soils and natural waters.

II. Fundamental Principles

A. The Excitation Process

The basic physical process of molecular luminescence spectroscopy begins with the transfer of energy from electromagnetic radiation to electrons in a molecule of interest. These excited electrons are transferred from ground-state molecular orbitals to excited-state nonbonding or antibonding orbitals. A typical transition, is the promotion of π electrons to π^* orbitals in aromatic C bonds. In this case the transfer is from a singlet ground state[1] (S_0) to an excited-singlet state (S_1, S_2, S_3, etc.). The transition of a ground-state singlet electron to an excited-triplet state ($S_0 \rightarrow T_1$) is a spin-forbidden transition and has a low probability for occurrence. The $S_0 \rightarrow S_i$ transition, where $i > 2$, occurs on a time scale of 10^{-15} sec. This time is significantly less than that required for a single molecular vibration; thus the initial excited-state molecule maintains the same geometry of bond angles and bond distances as that of the ground-state species. Since the most stable geometry of the excited state is often different from that of the ground state, the S_i state will be both an excited electronic state and an excited vibrational state. The vertical transition from $S_0 \rightarrow S_i$, known as a Frank Condon transition, is depicted in Figure 1 for a single-bond coordinate of a simple hypothetical molecule. The horizontal lines in Figure 1 represent vibrational quantum states of the S_0 and S_i electronic states and the curves, the potential energy wells of the S_0 and S_i electronic states. For real molecules the presence of multiple bonds results in complex multidimensional patterns of energy levels.

By an elapsed time of 10^{-13} sec, the first excited state relaxes through molecular vibration and drops down to the lowest vibrational level of the same excited state. The excited molecule subsequently releases energy to other vibrational modes or through collisions with solvent molecules arriving at the lowest vibrational level of the lowest excited electronic state, S_1, within approximately 10^{-11} sec. This relaxation process from S_i to S_1 is known as internal conversion.

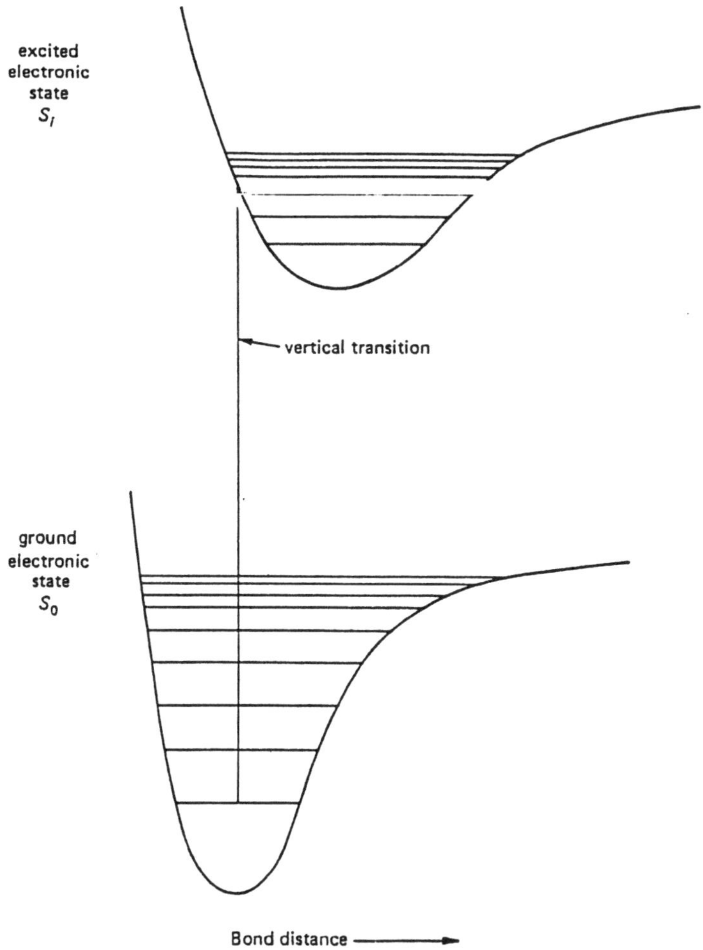

Figure 1. Frank–Condon electronic transition for a hypothetical molecule. (Reprinted with permission of Macmillan Publishing Company, from *Introduction to Organic Chemistry*, 3rd Ed., by A. Streitwieser, Jr. and C.H. Heathcock. Copyright © 1985 by Macmillan Publishing Company.)

The S_1 state commonly has a lifetime on the order of 10^{-9} sec. Relaxation from the S_1 state can occur by four different mechanisms (Fig. 2):

1. S_1 can emit a photon and undergo an electronic transition to the ground state. This is the process of fluorescence. Fluorescence always occurs at longer wavelengths than the original exciting radiation. This is a direct result of the energy lost during the short-lived internal conversion processes.
2. S_1 can continue to relax through internal vibrations or through collisions with solvent or solute molecules. This is again termed internal conversion and is a nonradiative process.

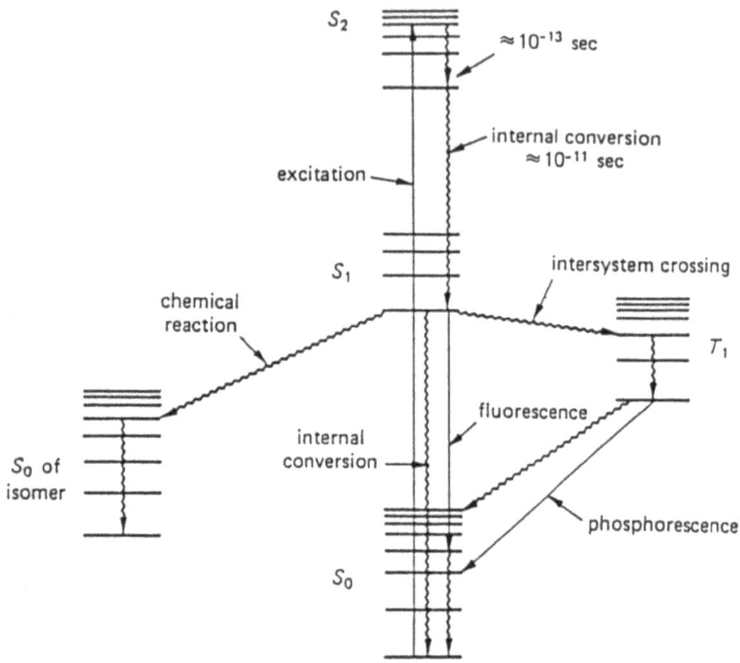

Figure 2. Jablonski diagram for a hypothetical molecule. The wavy lines indicate non-radiative processes. (Reprinted with permission of Macmillan Publishing Company, from *Introduction to Organic Chemistry*, 3rd Ed., by A. Streitwieser, Jr. and C.H. Heathcock. Copyright © 1985 by Macmillan Publishing Company.)

3. S_1 can undergo internal conversion to an excited vibrational level of the ground state of a different chemical compound, or combine with another molecule to form a polymerized compound. This is a photochemical transformation and is also a nonradiative process.
4. S_1 can undergo intersystem crossing to the triplet state, T_1. Recall that a triplet state is one in which one electron spin has been changed so that the molecule has two unpaired electrons that cannot repair without a second spin reversal. Even though intersystem crossing is forbidden by quantum mechanic selection rules, it still has some finite probability of occurring.

The T_1 molecule has a much longer lifetime than its S_1 precursor; lifetimes $>10^{-5}$ sec are not uncommon. Relaxation from this state can occur through the emission of a photon resulting in phosphorescence, or by nonradiant relaxation processes similar to the ones discussed for S_1 molecules.

B. Fluorescence Measurements

The luminescence properties that can be measured for a given experimental system are as follows:

1. Fluorescence emission spectra.
2. Fluorescence excitation spectra.
3. Fluorescence polarization.
4. Fluorescence quenching efficiency.
5. Fluorescence lifetimes and decay rates.

Additional parameters, such as synchronous luminescence and phosphorescence can also be measured, but they are not discussed in this paper. The five properties listed here are all sensitive to the specific chemical environment that is "seen" by a given fluorophore. Comparison of these properties measured in interfacial regions versus homogeneous solutions, provides the surface scientist with much information about the colloid–solution interface.

C. The Experiment

Often, minimal sample preparation is required for fluorescence measurements. Dilute colloidal suspensions (≤ 1 g solid L^{-1}) can be placed directly into quartz cuvettes, which in turn are put into the spectrophotometer. Whereas more dense colloidal suspensions can sometimes be used, increases in suspension density result in increases in light scattering and light absorption and a subsequent decrease in the signal-to-noise ratio. Provisions can be made to control sample temperatures through the use of thermostated cell holders. This is particularly useful in static quenching studies as discussed later. Cuvette stirrers can be added to most commercial fluorescence spectrophotometers to allow studies of unstable colloidal suspensions.

Steady-state fluorescence measurements can be made with a fluorescence spectrophotometer (spectrofluorometer). The optical diagram of a commercial spectrofluorometer is shown in Figure 3. Excitation energy is provided by a high-energy continuum light source, such as a Xe lamp. Incident radiation is passed through an excitation monochronometer and into a sample cuvette. Radiation emitted by the sample is monitored at an angle 90° to the excitation radiation, to minimize signal distortion due to light scattering. Wavelength selection of the emitted radiation is accomplished with an emission monochronometer, after which the emitted light is measured by the detector. In most spectrofluorometers, signal detection is achieved with a photomultiplier tube. Excitation spectra are collected by maintaining a constant emission wavelength while varying the setting of the excitation monochronometer, and emission spectra are measured by varying the setting of the emission monochronometer, for a constant excitation wavelength.

Polarized fluorescence measurements can be obtained by placing a polarization filter between the excitation monochronometer and the sample, and between the sample and the emission monochronometer.

Time-resolved fluorescence decay measurements are made by monitoring sample emission while exciting the sample with a pulsed or modulated source of excitation radiation. The emitted light is directed to either a rapid-detecting photomultiplier, a photodiode array, or a single photon counter, to allow for the collection of picosecond to nanosecond time scale information.

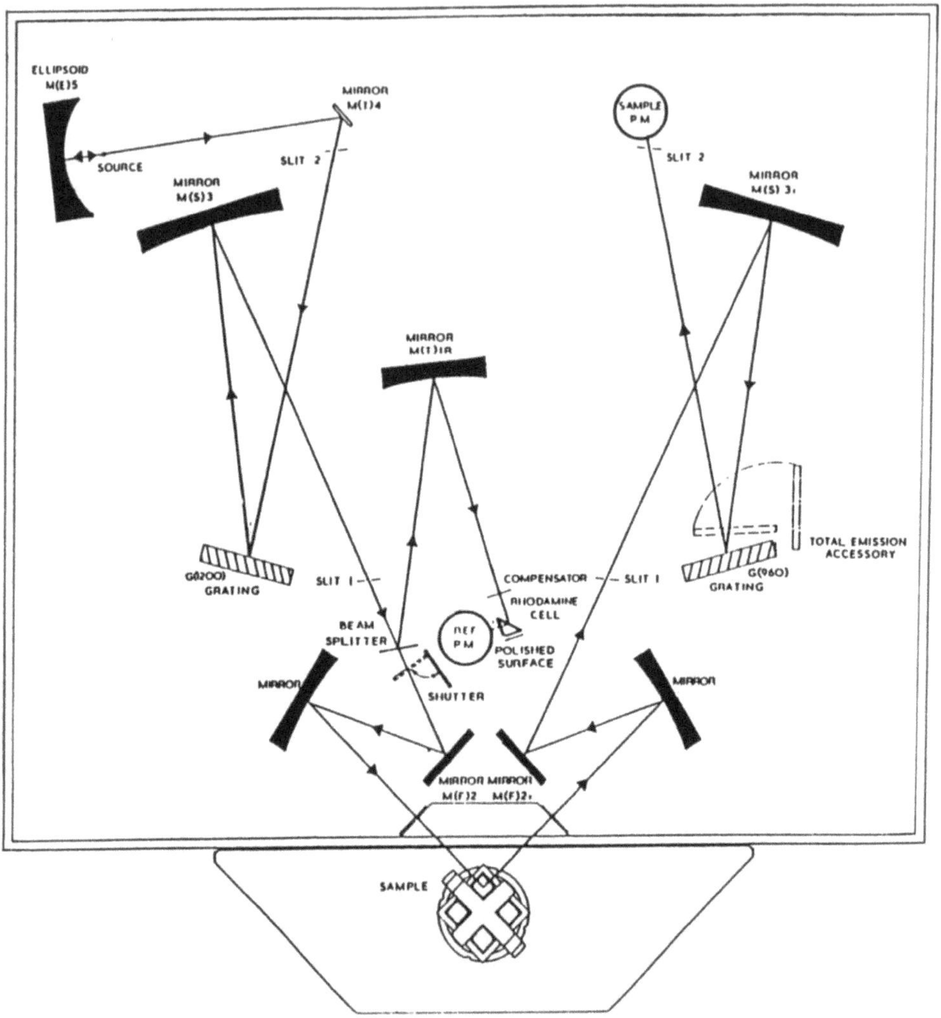

Figure 3. Optical diagram for a Perkin-Elmer LS-5 Spectrofluorometer (Perkin-Elmer Corp.).

D. Effects of the Chemical Environment on Molecular Fluorescence

1. Solvent Effects

The changes which take place in the electron distribution of a molecule when it undergoes transition from the S_0 state to the S_1 state, can cause changes in dipolar and hydrogen-bonding properties of the solute. If the fluorophore is more polar in the excited state than in the ground state, fluorescence will occur at longer wavelengths in polar solvents than in nonpolar solvents. This is a result of

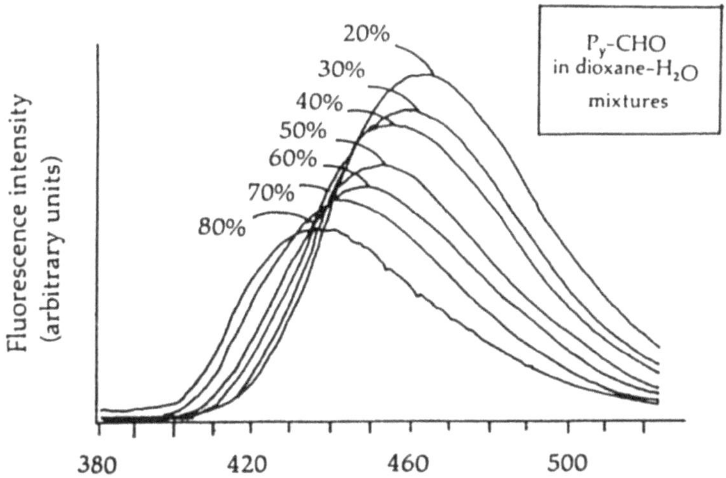

Figure 4. Solvent dependence of the monomer fluorescence of pyrene-3-carboxaldehyde in various dioxane–water mixtures (percentage of dioxane is given). (Reprinted with permission from J.K. Thomas, "The chemistry of excitation at interfaces," ACS Monograph 181. Copyright 1984 American Chemical Society.)

stabilization of the excited-state species, relative to the ground state. Such behavior is typical of most $\pi \rightarrow \pi^*$ transitions (Schulman, 1977) and is illustrated in Figure 4. In this study (Thomas, 1984) 1-pyrenecarboxaldehyde was dissolved in solutions of 1,4-dioxane and H_2O, containing from 20 to 80% 1,4-dioxane. This provided a range in solvent polarity (since the dielectric constant of H_2O is 78.54 and that of 1,4-dioxane is 2.21), with the most polar solution containing 20% dioxane and the least polar 80% dioxane. Decreases in the polarity of the mixed solvent solutions (from 20% dioxane to 80% dioxane) resulted in decreases in the total fluorescence intensity and in the wavelength corresponding to the maximum fluorescence emission (Fig. 4), indicating that the excited state 1-pyrenecarboxaldehyde molecule is more polar than the ground-state species. Variations in solvent polarity can also result in changes in fluorescence fine structure. Most notable is the dependence of the first and third vibronic transitions in the pyrene molecule (Thomas, 1984). The lowest vibrational band (the O-O) band is a forbidden vibration and is strongly affected by the surrounding solvent. In contrast the third vibrational band of pyrene is strong and is an allowed vibration which shows little dependence on solvent polarity. Thus the ratio of the first and third peaks, in the emission spectrum of pyrene can be used as an indicator of relative solvent polarity. Figure 5 shows the effect of the dielectric constant of the solvent on the fluorescence fine structure of pyrene. As the dielectric constant is increased from 1.89 (hexane) to 36.01 (acetonitrile) the ratio III/I decreases, where III and I are the intensities of the third and first vibronic emission peaks of pyrene, respectively.

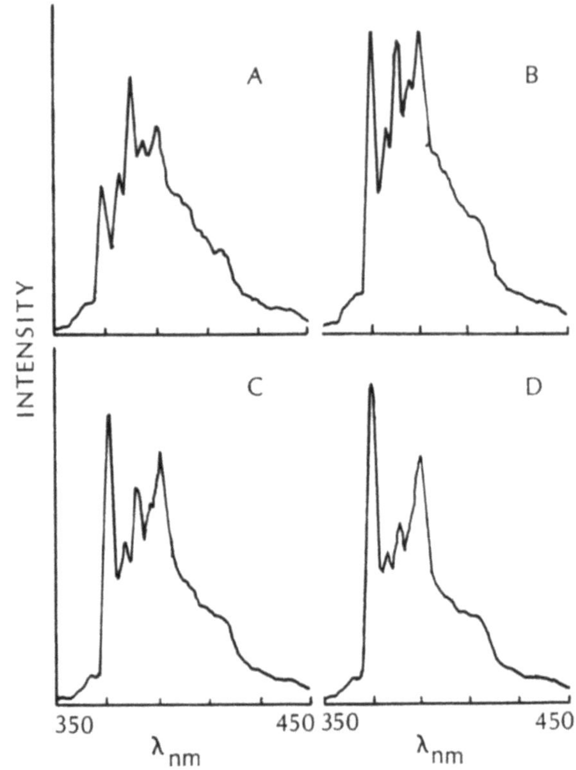

Figure 5. Solvent dependence of vibronic band intensities of pyrene monomer fluorescence: (A) n-hexane, III/I = 0.98, ε = 1.9; (B) n-butanol, III/I = 1.65, ε = 17.1; (C) methanol, III/I = 0.75, ε = 32.6; and (D) acetonitrile, III/I = 0.75, ε = 36.0. (Reprinted with permission from J.K. Thomas, "The chemistry of excitation at interfaces," ACS Monograph 181. Copyright 1984 American Chemical Society.) The solvent dielectric constants (ε) are from Weast and Astle (1980).

2. Molecular Aggregation

At higher solute concentrations, there is a tendency for molecules to form aggregates in both ground and excited states. The close approach of two conjugated molecules often results in an interaction of their *p* electron systems. This can cause a destructive or constructive addition of dipole moments of the monomers in both the S_0 and S_1 states, producing a change in fluorescence spectra. When planer aggregation occurs, the dipole moments of the S_1 molecules generally reinforce one another, resulting in a shift in fluorescence to longer wavelengths. If aggregation causes an overlap of the aromatic rings, the excited-state dipole moments will tend to cancel, causing excitation to S_1 to be somewhat forbidden. This latter form of aggregation generally results in intersystem crossing to the T_1, triplet state, where phosphorescence is favored over fluorescence in rigid media.

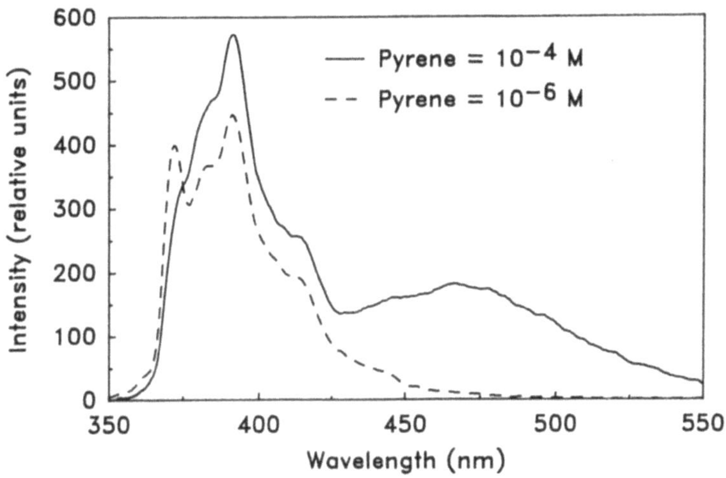

Figure 6. Effect of pyrene concentration on monomer and excimer fluorescence, in methanol. Samples were excited at 272 nm.

In some instances, an excited solute molecule can combine with a ground-state, solute molecule and produce an emitting dimer. When such a complex is composed of the same type of solute, it is called an excimer; when it is composed of two different types of molecules it is known as an exciplex. The spectral properties of excimers and exciplexes are typically different from those of S_1 monomers. Excimer formation in pyrene results in a broad featureless emission centered on the wavelength region of approximately 460 to 480 nm. Figure 6 shows the fluorescence emission spectra of pyrene in methanol. As the total pyrene concentration is increased from 10^{-6} to 10^{-4} mol L^{-1} a broad diffuse peak (centered at approximately 475 nm) indicative of pyrene excimer emission becomes apparent. Thus the appearance of excimer emission provided direct evidence of association between two or more pyrene molecules in solution. Concominant with excimer emission is a decrease in the emission of the 0-0 vibrational band.

3. Fluorescence Quenching

The presence of cosolutes can strongly alter the fluorescence behavior of an excited state fluorophore. As indicated, this can result in the formation of an exciplex which has different spectral properties from the free fluorophore molecule. Cosolutes can also greatly reduce the amount of light emitted by an excited-state fluorophore by the process of quenching. Quenching is a generic term used to describe the nonradiant transfer of energy from an excited-state species to other molecules in the solution. This relaxation process can be subdivided into two broad categories of state quenching and dynamic quenching. In the former case, a ground-state complex is formed between the fluorophore and other solute molecules. Upon excitation, the fluorophore transfers energy to the other molecule(s) in the complex, which then relaxes through a nonradiant process. Thus, the fluorescence quantum yield of an excited-state molecule, in the presence of a static quencher, is controlled by the strength (thermodynamics) of the ground-state complex (Schulman, 1977). Dynamic quenching is a diffusion-controlled process in which an excited fluorophore collides with a ground-state solute molecule. Again, relaxation of the excited quencher occurs through a nonradiant process. The efficiency of a dynamic quencher is limited by the fluorescence lifetime of the excited state of the fluorophore and by the bimolecular diffusion rates of the fluorophore and the quencher in the bulk solvent. This can be described as follows:

Process	*Reaction*	*Rate Parameter*
Fluorescence	$M^* \rightarrow M + h\nu$	k_F
Internal conversion	$M^* \rightarrow M$	k_G
Intersystem crossing	$M^* \rightarrow M$	k_T
Quenching	$M^* + Q \rightarrow M + Q$	$k_Q[Q]$

where M^* and M are the excited and ground states of the fluorophore, and Q is the quencher (Lumb, 1978). The fluorescence quantum yield in the absence of the quencher, Φ_0, is given by

$$\Phi_0 = \frac{k_F}{k_F + k_G + k_T}$$

The fluorescence quantum yield in the presence of the quencher, Φ_q, can be written as

$$\Phi_q = \frac{k_F}{k_F + k_G + k_T + k_Q[Q]}$$

where $[Q]$ is the concentration of the quencher. Thus, the relative quantum yield, Φ_0/Φ_q is given by

$$\frac{\Phi_0}{\Phi_q} = \frac{k_F(k_F + k_G + k_T + k_Q[Q])}{k_F(k_F + k_G + k_T)} \tag{1}$$

$$= 1 + \frac{k_Q[Q]}{k_F + k_G + k_T} \tag{2}$$

$$= 1 + \tau_M k_Q[Q] \tag{3}$$

where τ_M is the fluorescence lifetime in the absence of Q (Lumb, 1978). Equation 3 is a form of the well-known Stern–Volmer equation. The Stern–Volmer quenching constant, K_{SV} is equal to the product $\tau_M k_Q$. In the case of a diffusion controlled quenching process, k_Q is equal to k_{diff} and Equation 3 can be written as

$$= 1 + \tau_M k_{\text{diff}} [Q] \qquad (4)$$

When applied to conditions of static quenching, K_{SV} is equal to the conditional association constant, ${}^cK$ for the formation of the complex between the ground state fluorophore and quencher (MQ) which is given by

$${}^cK = \frac{[MQ]}{[M][Q]} \qquad (5)$$

In practice, the ratio I_0/I is often substituted for the relative quantum yield, where I_0 is the fluorescence intensity of a solution of fluorophore M, measured at a designated excitation and emission wavelength (λ_{ex} and λ_{em}), in the absence of Q, and I is the fluorescence intensity of M, measured at λ_{ex} and λ_{em}, in the presence of Q.

$$\frac{I_0}{I} = 1 + {}^cK[Q] \qquad (6)$$

A distinction between static and dynamic quenching can be obtained from an analysis of fluorescence lifetimes. Since dynamic quenching occurs as a result of collisions between an excited-state fluorophore and a quencher molecule, it results in a decrease in the observed fluorescence lifetime. In contrast, fluorescence lifetimes are not altered in systems that undergo static quenching (Ingle and Crouch, 1988).

III. Applications of Fluorescence to Colloidal Systems

A. Potential Probe Molecules

A variety of different molecules have been used as fluorescence probes of interfacial regions in colloidal systems (Krenske et al., 1980; Bergaya and Van Damme, 1983; DellaGuardia and Thomas, 1983; Schoonheydt et al., 1986). Because its spectral properties are easily influenced by the surrounding environment, pyrene and various cationic pyrenyl derivatives, such as [3-(1-pyrenyl)propyl]trimethylammonium, are commonly used as probes in suspensions 2:1 layered aluminosilicate clays. Pyrene will not adsorb onto charged clay particles in aqueous solutions to any appreciable degree, but it will adsorb onto clay minerals in the presence of nonaqueous solvents or coadsorbed surfactants. Anionic pyrene derivatives, such as pyrenesulfonic acid, can be used to study positively charged oxide surfaces. Nonpolar arenes adsorb onto a number of organic surfaces, and they can be used in studies of colloidal soil organic matter.

In choosing the appropriate probe molecules for a given interfacial environment, one must consider both the mechanism of adsorption of a fluorophore molecule by the surface of interest and the specific aspects of the interface that are to be probed. Some molecules report back on the local surface acidity, some on the apparent interfacial solvent polarity, and some on their relative mobility in the interfacial region. The reader is referred to the reviews of Thomas (1984, 1987) for a more thorough list of potential fluorescent probes.

B. Structural Quenching

An important aspect of fluorescence studies in clay mineral suspensions is the attenuation of the fluorescence of adsorbed fluorophores by quenching cations located within the crystal structure of clay minerals. Most common is the quenching of excited-state adsorbates by octahedral Fe(III) in 2:1 layered aluminosilicates. This phenomenon can present a serious limitation to the use of fluorescence spectroscopy in many natural systems of interest. Most if not all soil clay minerals, and most reference clay minerals contain appreciable quantities of isomorphically substituted Fe (Dixon and Weed, 1977), rendering the relative fluorescence quantum yields of many probe molecules below the practical detection limits of most conventional instrumentation. In practice this limitation is overcome by the use of reference clay minerals with low Fe content such as hectorite.

Several investigators have reported significant quenching of adsorbed fluorophores by octahedral Fe(III) in 2:1 layered aluminosilicates (Krenske et al., 1980; Bergaya and Van Damme, 1983; DellaGuardia and Thomas, 1983; Schoonheydt et al., 1986). This phenomenon was well documented by Krenske et al. (1980), in a study of the photophysics of tris(2,2′-bipyridine)Ru(III) and tris(2,2′-bipyridine)Cr(III) adsorbed onto Na-saturated Wyoming montmorillonite, hectorite, and a synthetic montmorillonite, which contained Cr(III) in the octahedral layer. The quenching efficiencies of the structural quenchers located within the clay mineral lattices were compared to those of Cr(III), Eu(III), and Fe(III) associated with cation exchange sites on the mineral surfaces. The greatest quantum yields for both fluorescent molecules were measured in suspensions of Na-hectorite (which contained the lowest quantity of structural quenching cations). Both fluorophores were more strongly quenched on the Na-montmorillonites. Octahedral Fe(III) and Cr(III) were found to be more effective than exchangeable Fe(III) and Cr(III) at quenching the fluorescence of adsorbed tris(2,2′-bipyridine)Ru(III) and tris(2,2′- bipyridine)Cr(III). Krinske et al. (1980) suggested that the adsorbed fluorophores were closely associated with siloxane ditrigonal cavities on the clay surfaces, and that the average distance between an excited-state fluorophore and a quencher was less if the quenching cation was located in an octahedral lattice site rather than on a cation exchange site. This shorter distance resulted in greater fluorescence quenching (Krenske, et al. 1980). A similar mechanism has been invoked by Wheeler and Thomas (1988) to describe the greater quenching efficiency of octahedral Cu(II), relative to exchangeable Cu(II), in a fluorescence study of (4-(1-Pyrenyl)butyl)trimethylammonium bromide adsorbed onto synthetic hectorites containing from 0 to approximately 3% (by oven dried weight) octahedral Cu(II).

The phenomenon of structural quenching can be used to measure conditional association constants for the binding of fluorescent solutes to organic macromolecules. Gauthier et al. (1986) used a fluorescence quenching method to measure the binding of pyrene, phenanthrene, and anthracene by humic and fulvic acids, in aqueous solutions. In this system, a nonfluorescent, ground-state complex is formed between the polycyclic aromatic hydrocarbons (PAH) and the anionic, polymeric, organic macromolecules. This complex formation is thought to be due to van der Waals attractions between the conjugated *p* bonds in the PAH and similar aromatic structures found in the "hydrophobic" portions of the humic and fulvic acids. Equation 6 was used to calculate values of ^cK for the formation of PAH–humic acid and PAH–fulvic acid complexes (Gauthier et al., 1986). Traina et al. (1989) used this technique to study the effects of counterion valence on the binding of naphthalene to water-soluble organic carbon (WSOC) (a mixture of humic and fulvic acids and lower-molecular-weight organic molecules). Decreases in the intrinsic fluorescence of naphthalene, resulting from additions of humic and fulvic acid, were used to calculate values of ^cK, in $NaClO_4$, $Ca(ClO_4)_2$, and $Al(ClO_4)_3$ solutions. Whereas in the absence WSOC, variation in cation type did not influence the fluorescence quantum yield of naphthalene, the fluorescence quenching efficiency of WSOC (and thus the values of ^cK) decreased in the order of Na > Ca > Al. Additions of di- or trivalent cations (Ca^{2+}, Al^{3+}) coagulated the macromolecules present in WSOC, decreasing the molar volume of the "hydrophobic" regions within polymeric organic solutes. This more compact colloidal configuration decreased the binding of naphthalene by the WSOC (Traina et al., 1989).

C. Polarity of the Interfacial Environment

The presence of discrete, charged surface sites on hydrous metal oxides, layered aluminosilicates, and natural and synthetic organic colloids can produce interfacial environments that have apparent solvent polarities greater than those found in the aqueous bulk solutions. These surfaces also contain regions that are more nonpolar than aqueous solutions. Judiciously chosen probe molecules can provide insights into the presence of both polar and nonpolar regions on the colloid surfaces. The emission spectrum of pyrene adsorbed onto air-dried, Na-saturated laponite (a synthetic hectorite) exhibits a value for the ratio of the first and third vibronic peaks (I/III) of 0.90, significantly lower than the value of 1.74 observed for pyrene dissolved H_2O (Labbé and Reverdy, 1988). Labbé and Reverdy (1988) attributed this to the adsorption of pyrene onto hydrophobic regions of the siloxanic oxygen planes of Na–laponite. Upon further dehydration under vacuum, the I/III ratio for adsorbed pyrene increases to a value of 1.80, indicating the presence of pyrene in a much more polar environments. Presumably this results from the partial removal of H_2O from the hydration shell of the adsorbed Na^+ ions and the subsequent interaction of the vicinal pyrene with the desolvated cation (Labbe and Reverdy, 1988).

Analyses of the ratios of the fluorescence intensities for the I/III vibronic bands of pyrene, [3-(1-pyrenyl)propyl]trimethylammonium (PN3), (1-pyrenyl)-trimethylammonium (PN), and (8-(1-pyrenyl)octyl)ammonium (PN8) ions adsorbed onto laponite, in both aqueous and methanol solutions, have shown the charged regions of the interlamellar surfaces of clay particles to be highly polar environments (Nakamura and Thomas, 1986; Viaene et al., 1987; Viaene et al., 1988). When sufficient quantities of alkylammonium surfactant molecules are adsorbed by clay mineral surfaces, the ratios of the I/III fluorescence vibrations of pyrenyl probe molecules decrease, indicating that the fluorescence probes are residing in less polar environments and are strongly associated with the adsorbed surfactants (Nakamura and Thomas, 1986; Viaene et al., 1987; Viaene et al., 1988).

Measurements of apparent solvent polarities are not restricted to inorganic colloidal systems. The relative polarities of organic colloids and micelles have been measured with fluorescence probes (Thomas, 1984; Turro and Kuo, 1985). Diagnostic changes in the first and third vibronic bands of pyrene, and changes in the value of λ_{max} (the emission wavelength of maximum intensity) for indoles, pyrenecarboxaldehyde, and other probe molecules, have provided insight into the spatial distributions of polar and nonpolar regions within many different organic micelles. Changes in spectra have also been used to ascertain the points of solubilization of specific probe molecules within micellular structures. Turro and Kuo (1985) used three fluorescence probes (pyrenecaroxaldehyde, PA; pyrene, P; and an indole detergent, In); to probe the microscopic characteristics of nonionic micelles (Triton series, C_9PhE_n). These probes can be listed in order of decreasing polarity according to the sequence PA $>$ P $>$ In. Thus, one would anticipate that in a micellular system containing a range of solvent polarities, PA would be located in the region of highest solvent polarity, P would be in an intermediate environment, and In would be found in the least polar region of the micelles. Each probe has a characteristic property that indicates the apparent polarity of the surrounding solvent environment. These characteristics are λ_F^{PA} and λ_F^{In}, the wavelength of maximum fluorescence emission for PA and In, respectively, and I/III, the intensity ratio of the first and third vibrational bands in the fluorescence spectrum of P. By comparing the fluorescence properties of each of these probe molecules in aqueous solutions containing micelles of triton to their fluorescence behavior in pure solutions of H_2O, Na–dodecylslufonate, n-butanol, and n-dodecane, Turro and Kuo (1985) were able qualitatively to describe the nature of the environment in which each probe resided. Experimental values of λ_F^{PA}, λ_F^{In}, and I/III indicated that PA resided at the exterior of the micelles, In was solubilized in the hydrophobic, nonpolar interior, and P was thought to reside in an intermediate location within the micelle.

D. Molecular Distributions, Configurations, Mobilities, and Interfacial Structures

The molecular distributions of cationic organic probes adsorbed onto clay surfaces can be studied with fluorescence spectroscopy. Adsorbed [4-(1-pyrenyl)-butyl]trimethylammonium (PN4), PN3, and PN8 cations tend to cluster on the surfaces of clay minerals (suspended in aqueous solutions) and exhibit excimer fluorescence at total concentrations (based on the volume of the entire suspension) below those at which excimer fluorescence occurs in homogeneous aqueous

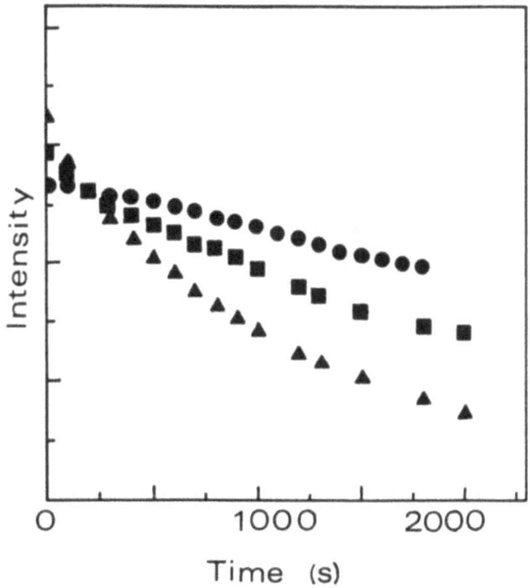

Figure 7. Time-dependent excimer fluorescence: $[Ca^{2+}] = 0$ M (●); $[Ca^{2+}] = 10^{-4}$ M (■); $[Ca^{2+}] = 10^{-3}$ M (▲). (Reprinted with permission from K. Viaene, J. Caigui, R.A. Schoonheydt, and F.C. De Schryver, "Study of the adsorption on clay particles by means of a fluorescent probe," *Langmuir* 3:107–111. Copyright 1987 American Chemical Society.)

solutions. Under these conditions, all excimer emission originates from adsorbed surface species (DellaGuardia and Thomas, 1983; Nakamura and Thomas, 1986; Viaene et al., 1987; Viaene et al., 1988). Analyses of excitation spectra and fluorescence lifetime measurements have shown that excimer emission (measured at λ_{em} = 480 nm) originates from ground-state complexes, consisting of two or more adsorbed probe molecules (DellaGuardia and Thomas, 1983; Viaene et al., 1987; Viaene et al., 1988). The probe molecule, PN, also tends to form ground-state complexes when adsorbed by clays, but this does not result in excimer emission (Viaene et al., 1988).

When observed on the time scales of fluorescence lifetime measurements (10^{-9} to 10^{-7} sec) surface clusters of adsorbed organic cations appear to be statically formed; yet reorientation of adsorbed surface species can occur over longer periods of time. Viaene et al. (1987) found a decrease in excimer emission from PN3 (measured at λ_{em} = 480 nm) over a time of 0 to 2000 sec, following the addition of the probe molecule to aqueous clay suspensions of Na-saturated hectorite, laponite, and montmorillonite. The addition of Ca^{2+} ions to these systems increased the excimer fluorescence intensities at time zero (after PN3 adsorption) and increased the rate of decrease in excimer emission. Both of these phenomena occurred in proportion to the total Ca^{2+} concentration (Fig. 7, Viaene et al., 1987). Viaene et al. (1987) proposed that the initial adsorption of PN3 occurs primarily on the exterior surfaces of the clay particles, resulting in high local concentrations of PN3 cations, and large excimer emission intensities.

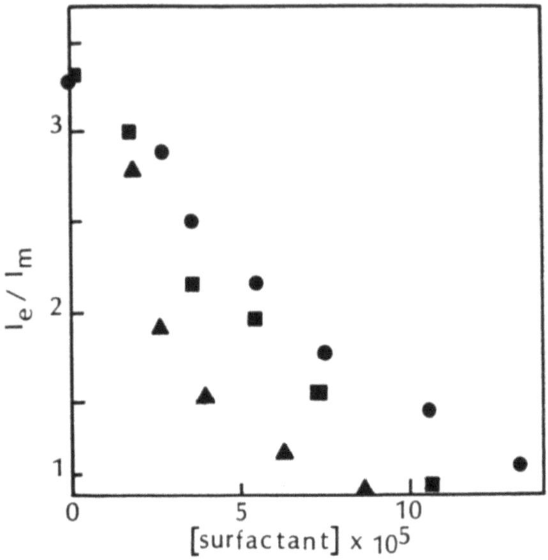

Figure 8. Influence of coadsorbed detergent molecules on excimer fluorescence: number of C = 14 (●); C = 16 (■); C = 18 (▲). (Reprinted with permission from K. Viaene, J. Caigui, R.A. Schoonheydt, and F.C. De Schryver, "Study of the adsorption on clay particles by means of a fluorescent probe," *Langmuir* 3:107–111. Copyright 1987 American Chemical Society.)

Excimer emission results from the static formation of ground-state complexes. Initially, excimer formation (at time zero) is enhanced by the presence of Ca^{2+} ions, which induce flocculation of the clay particles, restricting the availability of the interlamellar regions and increasing the concentrations of pyrenyl cations on the exterior surfaces immediately subsequent to adsorption. Following adsorption, PN3 cations migrate into the interlamellar regions of clay particles, decreasing the local concentration and thus the fluorescence intensity of excimer emissions. As migration into the interlamellar regions of the more flocculated Ca–clays proceeds, the mobilities of the pyrenyl cations are restricted (compared to Na–clays) and ground-state cluster formation is inhibited; thus the rate of decline of excimer emission is greater (Viaene et al., 1987).

The formation of ground-state clusters of pyrenyl cations on montmorillonite, hectorite, and laponite in aqueous solutions has been attributed to nonhomogeneous distributions of cation exchange sites on the mineral surfaces (Nakamura and Thomas 1985, 1986) and to the formation of "hydrophobic" micelles of adsorbed pyrenyl cations (Viaene et al., 1987). The latter mechanism was supported by the observation that excimer fluorescence is not observed from adsorbed PN3 and PN8 cations on laponite (indicating an unclustered distribution of the probe molecules on the clay surface) when the clays are suspended in a nonaqueous solvent such as methanol (Viaene et al., 1988). (Recall that these same cations exhibit strong excimer fluorescence in aqueous clay suspensions.)

Figure 9. Pyrene-labeled poly(acrylic acid). (Reprinted with permission from P. Chander, P. Somasundaran, N.J. Turro, and K.C. Waterman, "Excimer fluorescence determination of solid-liquid interfacial pyrene-labeled poly(acrylic acid) conformations," *Langmuir* 3:298–300. Copyright 1987 American Chemical Society.)

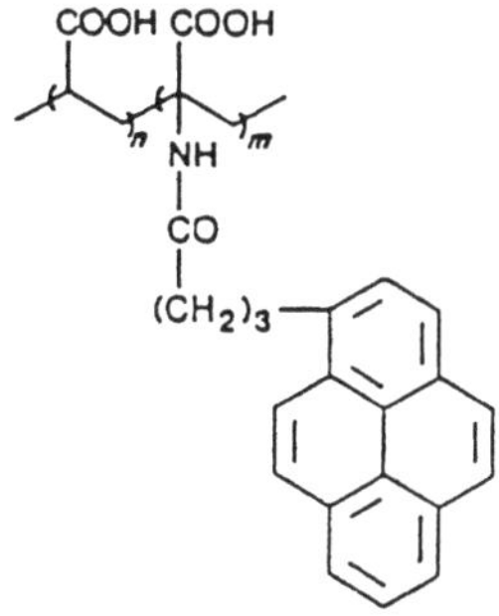

1

n = 0.985; m = 0.015;
DP = 510; MW = 40 000

The distribution of adsorbed pyrenyl cations can be altered by the presence of other organic cations on clay surfaces. Viaene et al. (1987) studied the effects of alkylammonium surfactants on the photophysics of adsorbed PN4. Adsorption of sufficient quantities of surfactants molecules dispersed the ground-state clusters of PN4 on the clay surface. This produced more homogeneous surface distributions of the pyrenyl cations and a corresponding reduction in excimer emission. This is illustrated by the data in Figure 8, which show a decrease in the ratio of PN4 excimer emission to PN4 monomer emission (I_E/I_M). For a given aqueous detergent concentration, increases in the chain length of the alkylammonium surfactant molecules (from C = 14 to C = 18) also produced a decrease in I_E/I_M (Fig. 8). This was again attributed to enhanced dispersal of the ground-state PN4 cations on the clay surface (Viaene et al., 1987). Similar results have been reported by DelaGuardia and Thomas (1983) and by Nakamura and Thomas (1986). DelaGuardia and Thomas (1983) also observed a decrease in the fluorescence anisotropy of adsorbed PN4 on clays containing coadsorbed alkylammonium surfactants. This was evidence of a reduction in the interactions of PN4 cations with the clay surface. Presumably, when adsorbed by clays that contain alkylammonium surfactants in the interfacial region, pyrenyl cations are "solubilized" into surfactant hemimicelles formed on the mineral surfaces. This results in an increase in the surface mobility of pyrenyl cations, which in turn causes a reduction in fluorescence anisotropy. As indicated by Viaene et al. (1987) the nature of the emission spectra of pyrenyl cations coadsorbed onto clays with alkylammonium surfactants can be used to infer the surface distribution of the surfactant hemimicelles. If the alkylammonium surfactants were present on the clay particles as a few large detergent clusters, then the local concentration (within each detergent cluster) of adsorbed pyrenyl cations would be high and excimer emission would occur. In contrast, the reduction in excimer emission which is commonly observed subsequent to the adsorption of alkylammonium surfactant molecules indicates that the detergent is present on the clay surfaces as numerous small clusters, each containing a locally small concentration of pyrenyl cations.

The effects of pH on the conformational structures of polymeric, polyelectrolytic macromolecules in aqueous solutions have been studied spectroscopically with pyrene-labeled poly(acrylic acid) (Py-PAA) (Fig. 9). In this system, the formation of pyrene excimers is enhanced by low pH values and minimized at high pH values (Turro and Arora, 1986). This is a direct result of pH-induced changes in the secondary and tertiary structures of the poly(acrylic) acid (PAA) polymer (predominantly coiled in acidic solutions, and linear in more basic solutions), which in turn alters the distance between two pyrene moieties covalently attached to the PAA "backbone." Chander et al. (1987) have extended this research in an elegant study of the effects of pH on the interfacial conformations of Py–PAA in aqueous suspensions of Linde A grade alumina (Al_2O_3). Figure 10a shows the effect of pH on the ratio of the monomer emission from Py–PAA (I_M) to the excimer emission (I_E) in solution and adsorbed to an aqueous solution of alumina. Although at pH values below 10, the value of I_M/I_E was greater for adsorbed Py–PAA than for dissolved Py–PAA, both sets of polymers exhibited a systematic decrease in excimer emission with increasing pH (Fig. 10a). This suggests that the general macromolecular structure of Py–PAA is not significantly altered by adsorption onto alumina. Figure 10b shows the fluorescence emission spectra of adsorbed Py–PAA at pH 4.9 and pH 7.1. It can clearly be seen that pH had no effect on monomer emission intensity, but the relative intensity of the excimer emission reduced at the more neutral pH. Chander et al. (1987) also examined the effect of changes in the composition of supporting electrolyte, subsequent to Py–PAA adsorption by alumina on the fluorescence spectra of Py–PAA. When Py–PAA was adsorbed onto alumina at pH 8 and the solids were separated from the aqueous phase and then resuspended in a pH 4.8 solution, no change in the pyrene excimer emission intensity was observed (Fig. 11). Once adsorbed in the uncoiled state, the polymers remained uncoiled, independent of the pH value of the surrounding solutions (Chander et al., 1987). In contrast, when Py–PAA was adsorbed onto alumina at pH 4.7 and the particles were then separated from the aqueous phase and resuspended in a pH 8 solution, some uncoiling of the adsorbed polymers took place, as evidenced by a decrease in excimer emission intensity (Fig. 11). Chander et al. (1987) suggested that the linear polymers were more strongly adsorbed by the alumina surface, resulting in a restriction in subsequent conformational changes. In contrast the coiled polymers were loosely bound by the surface of alumina and were able to undergo some conformational reorganization when exposed to different electrolyte solutions (Chander et al., 1987). Although it must be recognized that the conclusions of this study were based on the use of a single physical method – namely, fluorescence emission spectroscopy – and that confirmation of these results with one or more additional physical methods (such as two-dimensional nuclear magnetic resonance) is required, the results of Chander et al. (1987) show great promise in studies of molecular organization at interfacial regions. Indeed, these results may provide some insight into the organization of other polymers, such as humic and fulvic acids, at the mineral–solution interface.

E. Surface Acidity

Fluorophores that exhibit pH-dependent fluorescence spectra can be used to probe the relative surface acidities of colloid–solution interfaces. Figure 12

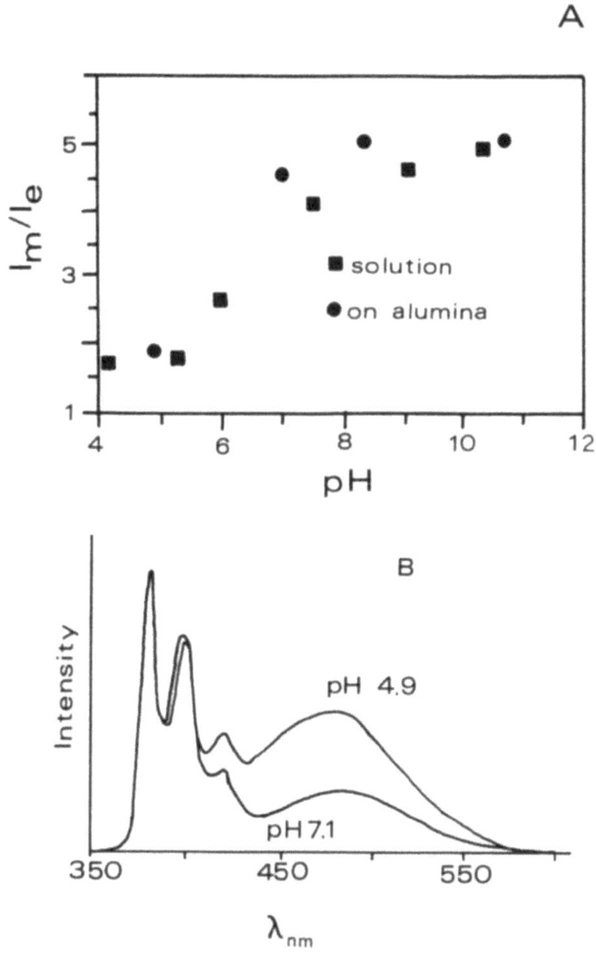

Figure 10. Effect of pH on the emission spectrum of pyrene-labeled poly(acrylic acid): (A) ratio of monomer to excimer emission intensity in aqueous solution and adsorbed onto alumina; (B) emission spectra of pyrene-labeled poly(acrylic acid) on Al_2O_3, at pH 4.9 and 7.1. (Reprinted with permission from P. Chander, P. Somasundaran, N.J. Turro, and K.C. Waterman, "Excimer fluorescence determination of solid-liquid interfacial pyrene-labeled poly(acrylic acid) conformations," *Langmuir* 3:298–300. Copyright 1987 American Chemical Society.)

shows the emission spectra of quinoline adsorbed onto Na–hectorite in the pH range of 3 to 9. The pK_a for quinoline is pH = 4.92 (Ainsworth et al., 1987); thus, at pH 8.82 virtually all of the quinoline in aqueous solutions should be in the molecular form. Yet the data in Figure 12 clearly indicate the presence of adsorbed quinolinium on the surface of Na–hectorite. Values of integrated fluorescence for the quinoline–hectorite suspensions (Fig. 12) and for pH 2.75, Na–acetate and pH 2.74, NaCl solutions that contained the same total quino-

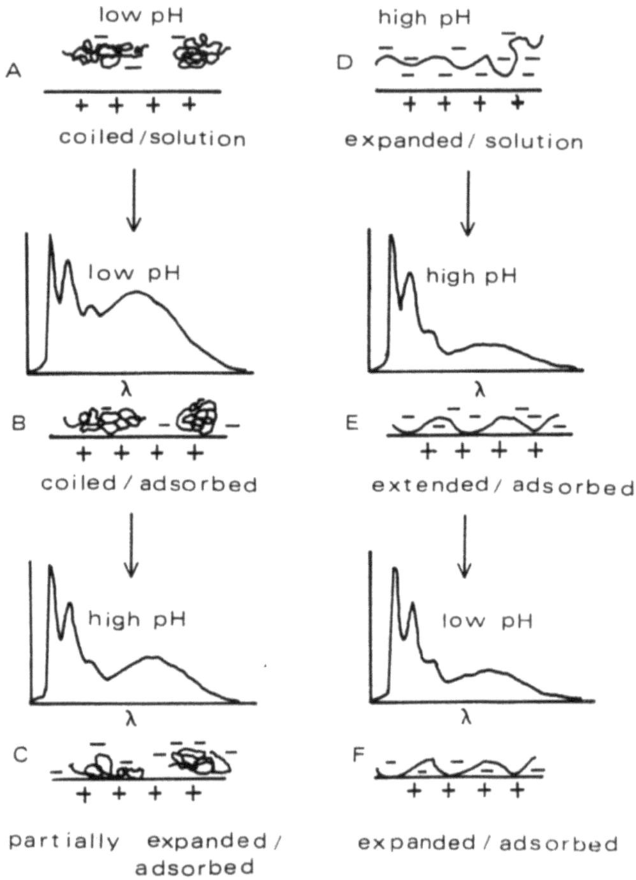

Figure 11. Schematic representation of the adsorption process of pyrene-labeled poly(acrylic acid) on alumina. (A) At low pH the polymer is coiled in solution, which leads to (B) adsorption in the coiled form. (C) Subsequent raising of the pH causes some expansion of the polymer. (D) Polymer at high pH in the solution is extended and binds (E) strongly to the surface in this conformation. (F) Subsequent lowering of the pH does not allow for sufficient interstrand interactions for coiling to occur. (Reprinted with permission from P. Chander, P. Somasundaran, N.J. Turro, and K.C. Waterman, "Excimer fluorescence determination of solid-liquid interfacial pyrene-labeled poly(acrylic acid) conformations," *Langmuir* 3:298–300. Copyright 1987 American Chemical Society.)

line concentrations as the hectorite suspensions are presented in Table 1. It is apparent that protonation of quinoline is promoted by the hectorite surface. This protonation may be a result of surface acidity, but it may also result from the preferential adsorption of the protonated quinolinium species directly from solution. The latter mechanism is supported by the greater fluorescence of quinoline in the presence of Na-acetate relative to NaCl (Table 1). Presumably some quinolinium is stabilized in solution by the formation of a quinolinium–acetate ion pair, resulting in greater emission intensity.

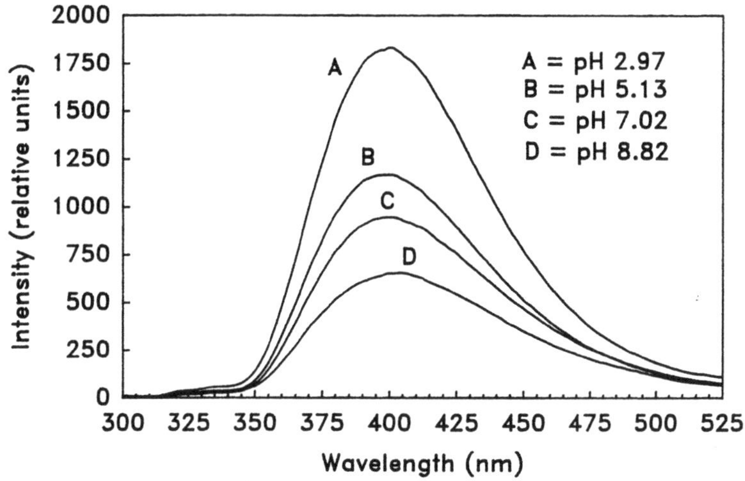

Figure 12. Effect of pH on the fluorescence emission spectrum of quinolinium adsorbed onto hectorite, in 0.02 M NaCl. The suspension pH values are presented.

Other molecules can also be used to study surface-enhanced protonation. Schoonheydt et al. (1986) observed the characteristic emission spectra of diprotonated proflavine, following the addition of the monoprotonated form, to suspensions of Na-saturated Camp Berteau montmorillonite (CB), Wyoming bentonite (WB), and Barasym SSM-100 (BS). Aminopyrene has been used to study the apparent surface acidities of colloidal silicas (Hite et al., 1986). The neutral aminopyrene molecule (AP) has a broad, featureless fluorescence emission centered at 440 nm in aqueous solution. In contrast, protonated aminopyrene (APH^+) exhibits a highly structured emission spectrum reminiscent of unsubstituted pyrene (Hite et al., 1986). Hite et al., (1986) have shown that aminopyrene adsorbed to colloidal silicas exhibits spectra similar to that of AP, APH^+, and spectra that

Table 1. Effect of pH, adsorption and anions on the integrated fluorescence emmission of quinolinium[a]

Sample matrix	pH	Integrated fluorescence
Hectorite[b]	2.97	294,384
Hectorite	5.13	192,838
Hectorite	7.02	162,329
Hectorite	8.82	117,072
Na–acetate	2.74	30,041
NaCl	2.75	27,568

[a] Ionic strength equals 0.05 mol l^{-1} for all samples.

[b] Quinolinium adsorbed onto hectorite in a matrix solution of 0.05 mol l^{-1} NaCl.

appear to be intergrades between these extremes. The APH^+-like spectra are found on the more acidic silicas and the AP-like spectra on the more basic silica surfaces Hite et al. (1986).

IV. Conclusions

As is true for other spectroscopies, there are some limitations in using fluorescence spectroscopy to study colloidal systems. Perhaps the most significant of these is the quenching of fluorescent adsorbates by structural Fe(III) located within the lattice of most 2:1 phyllosilicates. This restricts fluorescence studies of inorganic clay surfaces to those minerals that do not contain Fe(III). Additional sources of organic and inorganic quenchers (humic and fulvic acids, transition metals) may be present in suspensions of geologic and pedogenic clay minerals, increasing the difficulty of studying these materials.

As is discussed later, fluorescence spectroscopy is highly sensitive. This high sensitivity can in itself be an experimental limitation. Only trace quantities of a contaminating fluorophore (some molecule other than the chosen probe molecule) need be present in a colloidal system to be detected. If the fluorescence spectra of some contaminating solute or adsorbate overlaps with that of the probe molecule of interest, proper interpretation of the experimental data may be difficult.

Finally, fluorescence studies must obviously be restricted to systems containing fluorescent probes. While this may seem like a significant limitation, there are numerous types of potential probe molecules for the experimentalist to choose from.

In spite of these limitations, fluorescence spectroscopy offers many advantages to the colloid and surface chemist. Not the least of these is the ability to study aqueous collodial suspensions noninvasively. Incorporation of a flow-through cell in a spectrofluorometer allows one to easily study the kinetics of many chemical processes at the colloid–solution interface without the need to stop the reaction or resort to a separation of the solid and liquid phases.

The high analytical sensitivity of fluorescence spectroscopy allows one to study many adsorbates on colloidal surfaces, at concentrations approaching those found in natural environments. The absolute detection limit will differ depending on the fluorescence quantum yield of the specific fluorescent probe in use, but many fluorophores can be studied at extremely low solution or surface concentrations. Experiments conducted in the author's laboratory have studied the fluorescence spectra of pyrene coadsorbed with quinoline onto hectorite, at surface concentrations of approximately 10 nmol pyrene kg^{-1} of clay. In a separate experiment, solution studies of the fluorescence quenching of benzo(def)phenanthrene by humic and fulvic acids have been carried out at fluorophore concentrations of 60 to 100 pmol L^{-1} (S.J. Traina, unpublished data).

An additional strength of fluorescence spectroscopy, is the ability of one probe molecule to provide multiple types of information on the colloid–solution interface. This has been illustrated in the studies of pyrenyl cations on clay surfaces, where changes in the fluorescence lifetime, fluorescence fine structure (ratio of first and third vibronic peaks), and the relative contributions of monomer and excimer emissions to the total fluorescence quantum yield each provides unique, yet complimentary information on the environment in which the probe molecule resides.

Fluorescence and other luminescence spectroscopies can provide a wealth of information pertaining to the chemical environment of the colloid–solution interface. In practice, the researcher sends a judiciously chosen fluorescence probe into the interfacial region. The probe molecule then reports back to the spectroscopist on the chemical properties it encounters at the colloid surface. This report can include data on relative solvent polarities, molecular mobilities and distributions, chemical reaction rates, and surface configurations found at the colloid–solution interface.

By itself, fluorescence spectroscopy can provide new insight on the nature of the colloid–solution interface; however, no single spectroscopy can fully describe the interactions of a given molecule with other molecules or with a given surface. Comprehensive studies on chemical processes at the colloid–solution interface must include the use of complimentary techniques such as FTIR, NMR, and UV–VIS adsorption spectroscopies. A systematic application of these types of measurements to representative colloidal systems will assist soil scientists in their understanding of many surface-chemical phenomena.

Acknowledgments. The author thanks Dr. William Jaynes, Donna Spontak, and Blake Onken for their assistance in making some of the fluorescence measurements presented in this chapter. Additional recognition is given to the Ohio State University, and the Ohio Agricultural Research and Development Center, for their partial financial support of this research.

References

Ainsworth, C.C., J.M. Zachara, and R.L. Schmidt. 1987. Quinoline sorption on Na-montmorillonite: Contributions of the protonated and neutral species. Clays and Clay Min. 35:121–128.

Bergaya, F., and H. Van Damme. 1983. Luminescence of Eu^{3+} and Tb^{3+} ions adsorbed on hydrated layer-lattice silicate surfaces. J. Chem. Soc. Faraday Trans. 2. 79:505–518.

Chander, P., P. Somasundaran, N.J. Turro, and K.C. Waterman. 1987. Excimer fluorescence determination of solid-liquid interfacial pyrene-labeled poly(acrylic acid) conformations. Langmuir. 3:298–300.

DellaGuardia, R.A., and J.K. Thomas. 1983. Photoprocesses on colloidal clay systems. 2. Quenching studies and the effect of surfactants on the luminescent properties of pyrene and pyrene derivatives adsorbed on clay colloids. J. Phys. Chem. 87:3550–3557.

Dixon, J.B., and S.B. Weed. 1977. Minerals in soil environments. Soil Sci. Soc. Amer. Madison, Wisc.

Gauthier, T.D., E.C. Shane, W.F. Guerin, W.R. Seitz, and C.L. Grant. 1986. Fluorescence quenching method for determining equilibrium constants for polycyclic aromatic hydrocarbons binding to dissolved humic materials. Environ. Sci. Technol. 20:1162–1166.

Hite, P., R. Krasnansky, and J.K. Thomas. 1986. Spectroscopic investigations of surfaces using aminopyrene. J. Phys. Chem. 90:5795–5799.

Ingle, J.D., and Crouch, S.R. 1988. Spectrochemical Analysis. Prentice Hall, Englewood Cliffs, N.J.

Krenske, D., S. Abdo, H. Van Damme, M. Cruz, and J.J. Fripiat. 1980. Photochemical and photocatalytic properties of adsorbed organometallic compounds. 1. Luminescence quenching of tris(2,2'-bipyrine)ruthenium(II) and -chromium(III) in clay membranes. J. Phys. Chem. 84:2447–2457.

Labbé, P., and Reverdy, G. 1988. Adsorption characteristics of polycyclic aromatic compounds on clay: Pyrene as a photophysical probe on laponite. Langmuir. 4:419–425.

Lumb, M.D. 1978. Luminescence Spectroscopy. Academic Press, New York.

Nakamura, T., and J.K. Thomas. 1985. Photochemistry of materials adsorbed on clay systems. Effect of the nature of adsorption on the kinetic description of reactions. Langmuir. 1:568–573.

Nakamura, T., and J.K. Thomas. 1986. The interaction of alkylammonium salts with synthetic clays. A fluorescence and laser excitation study. J. Phys. Chem. 90:641–644.

Schoonheydt, R.A., J. Cenens, and F.C. De Schrijver. 1986. Spectroscopy of proflavine adsorbed on clays. J. Chem. Soc., Faraday Trans. 1, 82:281–289.

Schulman, S.G. 1977. Fluorescence and Phosphorescence Spectroscopy: Physicochemical Principles and Practice. Pergamon Press, New York.

Streitwieser, A., Jr., and C.H. Heathcock. 1985. Introduction to Organic Chemistry, 3rd Ed., Macmillan, New York.

Thomas, J.K. 1984. The chemistry of excitation at interfaces. ACS Monograph 181. American Chemical Society, Washington, D.C.

Thomas, J.K. 1987. Characterization of surfaces by excited states. J. Phys. Chem. 91:267–276

Traina, S.J., D.A. Spontak, and T.J. Logan. 1989. Effects of cations on complexation of naphthalene by water soluble organic carbon. J. Envir. Qual. 18:221–227.

Turro, N.J., and K.S. Arora. 1986. Pyrene as a photophysical probe for intermolecular interactions of water-soluble polymers in dilute solutions. Polymer 27:783–796.

Turro, N.J., and P. Kuo. 1985. Fluorescence probes for aqueous solutions of nonionic micelles. Langmuir 1:170–172.

Viaene, K., J. Caigui, R.A. Schoonheydt, and F.C. De Schryver. 1987. Study of the adsorption on clay particles by means of a fluorescent probe. Langmuir 3:107–111.

Viaene, K., R.A. Schoonheydt, M. Crutzen, B. Kunyima, and F.C. De Schryver. 1988. Study of the adsorption on clay particles by means of fluorescent probes. Langmuir 4:749–752.

Weast, R.C., and M.J. Astle. 1980. CRC Handbook of Chemistry and Physics, 61st ed. CRC Press, Boca Raton, Fla.

Wheeler, J., and J.K. Thomas. 1988. Quenching of excited (4-(1-pyrenyl)butyl)trimethylammonium bromide on synthetic hectorites containing lattice copper. Langmuir 4:543–546.

Index

MIX
Papier aus verantwortungsvollen Quellen
Paper from responsible sources
FSC® C105338

If you have any concerns about our products,
you can contact us on
ProductSafety@springernature.com

In case Publisher is established outside the EU,
the EU authorized representative is:
Springer Nature Customer Service Center GmbH
Europaplatz 3, 69115 Heidelberg, Germany

Printed by Libri Plureos GmbH
in Hamburg, Germany